TensorFlow 2 Reinforcement Learning Cookbook

Over 50 recipes to help you build, train, and deploy learning agents for real-world applications

Praveen Palanisamy

BIRMINGHAM—MUMBAI

TensorFlow 2 Reinforcement Learning Cookbook

Group Product Manager: Kunal Parikh

Publishing Product Manager: Aditi Gour

Senior Editor: David Sugarman

Content Development Editor: Joseph Sunil

Technical Editor: Manikandan Kurup

Copy Editor: Safis Editing

Project Coordinator: Aishwarya Mohan

Proofreader: Safis Editing

Indexer: Pratik Shirodkar

Production Designer: Alishon Mendonca

First published: January 2021

Production reference: 1140121

Published by Packt Publishing Ltd.

Livery Place

35 Livery Street

Birmingham

B3 2PB, UK.

ISBN 978-1-83898-254-6

www.packt.com

Packt.com

Subscribe to our online digital library for full access to over 7,000 books and videos, as well as industry leading tools to help you plan your personal development and advance your career. For more information, please visit our website.

Why subscribe?

- Spend less time learning and more time coding with practical eBooks and videos from over 4,000 industry professionals

- Improve your learning with Skill Plans built especially for you

- Get a free eBook or video every month

- Fully searchable for easy access to vital information

- Copy and paste, print, and bookmark content

Did you know that Packt offers eBook versions of every book published, with PDF and ePub files available? You can upgrade to the eBook version at packt.com and, as a print book customer, you are entitled to a discount on the eBook copy. Get in touch with us at customercare@packtpub.com for more details.

At www.packt.com, you can also read a collection of free technical articles, sign up for a range of free newsletters, and receive exclusive discounts and offers on Packt books and eBooks.

Contributors

About the author

Praveen Palanisamy works on advancing AI for autonomous systems as a senior AI engineer at Microsoft. In the past, he has developed AI algorithms for autonomous vehicles using deep reinforcement learning, and has worked with start-ups and in academia to build autonomous robots and intelligent systems. He is the inventor of more than 15 patents on learning-based AI systems. He is the author of *HOIAWOG: Hands-On Intelligent Agents with OpenAI Gym*, which provides a step-by-step guide to developing deep RL agents to solve complex problems from scratch. He has a master's in robotics from Carnegie Mellon University.

About the reviewer

Wilson Choo is a computer vision engineer involved in working on validating computer vision and deep learning algorithms on many different hardware configurations. His strongest skills include algorithm benchmarking, integration, app development, and test automation. He is also a machine learning and computer vision enthusiast. He often researches trending CVDL algorithms and applies them to solve modern-day problems. Aside from this, Wilson likes to participate in hackathons, where he showcases his ideas and competes with other developers. His favorite programming languages are Python and C++.

Packt is searching for authors like you

If you're interested in becoming an author for Packt, please visit authors. packtpub.com and apply today. We have worked with thousands of developers and tech professionals, just like you, to help them share their insight with the global tech community. You can make a general application, apply for a specific hot topic that we are recruiting an author for, or submit your own idea.

Table of Contents

2

Implementing Value-Based, Policy-Based, and Actor-Critic Deep RL Algorithms

3

Implementing Advanced RL Algorithms

4

Reinforcement Learning in the Real World – Building Cryptocurrency Trading Agents

5

Reinforcement Learning in the Real World – Building Stock/Share Trading Agents

6

Reinforcement Learning in the Real World – Building Intelligent Agents to Complete Your To-Dos

7

Deploying Deep RL Agents to the Cloud

8

Distributed Training for Accelerated Development of Deep RL Agents

9
Deploying Deep RL Agents on Multiple Platforms

Other Books You May Enjoy

Index

Preface

Deep reinforcement learning enables the building of intelligent agents, products, and services that can go beyond computer vision or perception to perform actions. TensorFlow 2.x is the latest major release of the most popular deep learning framework that is used to develop and train **deep neural networks (DNNs)**.

The book begins with an introduction to the fundamentals of deep reinforcement learning and the latest major version of TensorFlow 2.x. You'll then cover OpenAI Gym, model-based RL, and model-free RL, and learn how to develop basic agents. Moving on, you will discover how to implement advanced deep reinforcement learning algorithms such as actor-critic, deep deterministic policy gradients, deep-Q networks, proximal policy optimization, deep recurrent Q-networks, and the soft actor-critic algorithm to train your RL agents. You'll also explore reinforcement learning in the real world by building cryptocurrency trading agents, stock/share trading agents, and intelligent agents for automating task completion. Lastly, you will find out how to deploy deep reinforcement learning agents to the cloud and build cross-platform apps for the web, mobile, and other platforms using TensorFlow 2.x.

By the end of this cookbook, you will have gained a solid understanding of deep reinforcement learning algorithms with the help of easy-to-follow and concise implementations from scratch using TensorFlow 2.x.

Who this book is for

The book is for machine learning application developers, AI and applied AI researchers, data scientists, deep learning practitioners, and students with a basic understanding of the reinforcement learning concepts who want to build, train, and deploy their own reinforcement learning systems from scratch using TensorFlow 2.x.

What this book covers

Chapter 1, Developing Building Blocks for Deep Reinforcement Learning Using TensorFlow 2.x, provides recipes for getting started with RL environments, deep neural network-based RL agents, evolutionary neural agents, and other building blocks for both discrete and continuous action-space RL applications.

Chapter 2, Implementing Value-Based Policy Gradients and Actor-Critic Deep RL Algorithms, includes recipes for implementing value iteration-based learning agents and breaks down the implementation of several foundational algorithms in RL, such as Monte-Carlo control, SARSA and Q-learning, actor-critic, and policy gradient algorithms into simple steps.

Chapter 3, Implementing Advanced RL Algorithms, provides concise recipes to implement complete agent training systems using Deep Q-Network (DQN), Double and Dueling Deep Q-Network (DDQN, DDDQN), Deep Recurrent Q-Network (DRQN), Asynchronous Advantage Actor-Critic (A3C), Proximal Policy Optimization (PPO), and Deep Deterministic Policy Gradient (DDPG) RL algorithms.

Chapter 4, RL in the Real World – Building Cryptocurrency Trading Agents, shows how to implement and train a soft actor-critic agent in custom RL environments for bitcoin and ether trading using real market data from trading exchanges such as Gemini, containing both tabular and visual (image) state/observation and discrete and continuous action spaces.

Chapter 5, RL in the Real World – Building Stock/Share Trading Agents, covers how to train advanced RL agents to trade for profit in the stock market using visual price charts and/ or tabular ticket data and more in custom RL environments powered by real stock market exchange data.

Chapter 6, RL in the Real World – Building Intelligent Agents to Complete Your To-Dos, provides recipes to build, train, and test vision-based RL agents for completing tasks on the web to help you automate tasks such as clicking on pop-up/confirmation dialogs on web pages, logging into various websites, finding and booking the cheapest flight tickets for your travel, decluttering your email inbox, and like/share/retweeting posts on social media sites to engage with your followers.

Chapter 7, Deploying Deep RL Agents to the Cloud, contains recipes to equip you with tools and details to get ahead of the curve and build cloud-based Simulation-as-a-Service and Agent/Bot-as-a-Service programs using deep RL. Learn how to train RL agents using remote simulators running on the cloud, package runtime components of RL agents, and deploy deep RL agents to the cloud by deploying your own trading bot-as-a-service.

Chapter 8, Distributed Training for the Accelerated Development of Deep RL Agents,
contains recipes to speed up deep RL agent development using the distributed training
of deep neural network models by leveraging TensorFlow 2.x's capabilities. Learn how
to utilize multiple CPUs and GPUs both on a single machine as well as on a cluster of
machines to scale up/out your deep RL agent training and also learn how to leverage Ray,
Tune, and RLLib for large-scale accelerated training.

Chapter 9, Deploying Deep RL Agents on Multiple Platforms, provides customizable
templates that you can utilize for building and deploying your own deep RL applications
for your use cases. Learn how to export RL agent models for serving/deployment
in various production-ready formats, such as TensorFlow Lite, TensorFlow.js, and
ONNX, and learn how to leverage NVIDIA Triton or build your own solution to launch
production-ready, RL-based AI services. You will also deploy an RL agent in a mobile and
web app and learn how to deploy RL bots in your Node.js applications.

To get the most out of this book

The code in this book is extensively tested on Ubuntu 18.04 and Ubuntu 20.04 and should
work with later versions of Ubuntu if Python 3.6+ is available. With Python 3.6+ installed
along with the necessary Python packages, as listed at the start of each of the recipes, the
code should run fine on Windows and macOS X too.

Software/hardware covered in the book	OS requirements
Python 3.6 (or later versions)	Linux (any), Windows and macOS X
JavaScript	
HTML	

It is advised to create and use a Python virtual environment named tfrl-cookbook to
install the packages and run the code in this book. A Miniconda or Anaconda installation
for Python virtual environment management is recommended.

**If you are using the digital version of this book, we advise you to type the code yourself
or access the code via the GitHub repository (link available in the next section). Doing
so will help you avoid any potential errors related to the copying and pasting of code.**

It is highly recommended to star and fork the GitHub repository to receive updates
and improvements to the code recipes. We urge you to share what you build and
also engage with other readers and the community at https://github.com/
PacktPublishing/Tensorflow-2-Reinforcement-Learning-Cookbook/
discussions.

Download the example code files

You can download the example code files for this book from your account at www. packt.com. If you purchased this book elsewhere, you can visit www.packtpub.com/ support and register to have the files emailed directly to you.

You can download the code files by following these steps:

1. Log in or register at www.packt.com.
2. Select the **Support** tab.
3. Click on **Code Downloads**.
4. Enter the name of the book in the **Search** box and follow the onscreen instructions.

Once the file is downloaded, please make sure that you unzip or extract the folder using the latest version of:

- WinRAR/7-Zip for Windows
- Zipeg/iZip/UnRarX for Mac
- 7-Zip/PeaZip for Linux

The code bundle for the book is also hosted on GitHub at https://github.com/ PacktPublishing/Tensorflow-2-Reinforcement-Learning-Cookbook/. In case there's an update to the code, it will be updated on the existing GitHub repository.

We also have other code bundles from our rich catalog of books and videos available at https://github.com/PacktPublishing/. Check them out!

Download the color images

We also provide a PDF file that has color images of the screenshots/diagrams used in this book. You can download it here: https://static.packt-cdn.com/ downloads/9781838982546_ColorImages.pdf.

Conventions used

There are a number of text conventions used throughout this book.

Code in text: Indicates code words used in the recipes. Here is an example: "We will start with the implementation of the save method in the Actor class to export the Actor model to TensorFlow's SavedModel format."

A block of code is set as follows:

```
def save(self, model_dir: str, version: int = 1):
    actor_model_save_dir = os.path.join(model_dir, "actor",
str(version), "model.savedmodel")
    self.model.save(actor_model_save_dir, save_format="tf")
    print(f"Actor model saved at:{actor_model_save_dir}")
```

When we wish to draw your attention to a particular part of a code block, the relevant lines or items are set in bold:

```
if args.agent != "SAC":
    print(f"Unsupported Agent: {args.agent}. Using SAC Agent")
    args.agent = "SAC"
    # Create an instance of the Soft Actor-Critic Agent
    agent = SAC(env.observation_space.shape, env.action_space)
```

Any command-line input or output is written as follows:

```
(tfrl-cookbook)praveen@desktop:~/tensorflow2-reinforcement-
learning-cookbook/src/ch7-cloud-deploy-deep-rl-agents$ python
3_training_rl_agents_using_remote_sims.py
```

Bold: Indicates a new term, an important word, or words that you see onscreen. For example, words in menus or dialog boxes appear in the text like this. Here is an example: "Click on the **Open an Existing Project** option and you will see a popup asking you to choose the directory on your filesystem. Navigate to the Chapter 9 recipes and choose **9.2_rl_android_app**."

> Tips or important notes
> Appear like this.

Get in touch

Feedback from our readers is always welcome.

General feedback: If you have questions about any aspect of this book, mention the book title in the subject of your message and email us at customercare@packtpub.com.

Errata: Although we have taken every care to ensure the accuracy of our content, mistakes do happen. If you have found a mistake in this book, we would be grateful if you would report this to us. Please visit www.packtpub.com/support/errata, selecting your book, clicking on the Errata Submission Form link, and entering the details.

Piracy: If you come across any illegal copies of our works in any form on the Internet, we would be grateful if you would provide us with the location address or website name. Please contact us at copyright@packt.com with a link to the material.

If you are interested in becoming an author: If there is a topic that you have expertise in and you are interested in either writing or contributing to a book, please visit authors.packtpub.com.

Reviews

Please leave a review. Once you have read and used this book, why not leave a review on the site that you purchased it from? Potential readers can then see and use your unbiased opinion to make purchase decisions, we at Packt can understand what you think about our products, and our authors can see your feedback on their book. Thank you!

For more information about Packt, please visit packt.com.

1

Developing Building Blocks for Deep Reinforcement Learning Using Tensorflow 2.x

This chapter provides a practical and concrete description of the fundamentals of **Deep Reinforcement Learning (Deep RL)** filled with recipes for implementing the building blocks using the latest major version of **TensorFlow 2.x**. It includes recipes for getting started with RL environments, **OpenAI Gym**, developing neural network-based agents, and evolutionary neural agents for addressing applications with both discrete and continuous value spaces for Deep RL.

The following recipes are discussed in this chapter:

- Building an environment and reward mechanism for training RL agents
- Implementing neural network-based RL policies for discrete action spaces and decision-making problems
- Implementing neural network-based RL policies for continuous action spaces and continuous-control problems
- Working with OpenAI Gym for RL training environments
- Building a neural agent
- Building a neural evolutionary agent

Technical requirements

The code in the book has been extensively tested on Ubuntu 18.04 and Ubuntu 20.04 and should work with later versions of Ubuntu as long as Python 3.6+ is available. With Python 3.6 installed along with the necessary Python packages as listed before the start of each of the recipes, the code should run fine on Windows and macOS X too. It is advised to create and use a Python virtual environment named `tf2rl-cookbook` to install the packages and run the code in this book. Miniconda or Anaconda installation for Python virtual environment management is recommended.

The complete code for each recipe in this chapter will be available here: `https://github.com/PacktPublishing/Tensorflow-2-Reinforcement-Learning-Cookbook`.

Building an environment and reward mechanism for training RL agents

This recipe will walk you through the steps to build a **Gridworld** learning environment to train RL agents. Gridworld is a simple environment where the world is represented as a grid. Each location on the grid can be referred to as a cell. The goal of an agent in this environment is to find its way to the goal state in a grid like the one shown here:

Figure 1.1 – A screenshot of the Gridworld environment

The agent's location is represented by the blue cell in the grid, while the goal and a mine/bomb/obstacle's location is represented in the grid using green and red cells, respectively. The agent (blue cell) needs to find its way through the grid to reach the goal (green cell) without running over the mine/bomb (red cell).

Getting ready

To complete this recipe, you will first need to activate the tf2rl-cookbook Python/Conda virtual environment and pip install numpy gym. If the following import statements run without issues, you are ready to get started!

```
import copy
import sys
import gym
import numpy as np
```

Now we can begin.

How to do it...

To train RL agents, we need a learning environment that is akin to the datasets used in supervised learning. The learning environment is a simulator that provides the observation for the RL agent, supports a set of actions that the RL agent can perform by executing the actions, and returns the resultant/new observation as a result of the agent taking the action.

Perform the following steps to implement a Gridworld learning environment that represents a simple 2D map with colored cells representing the location of the agent, goal, mine/bomb/obstacle, wall, and empty space on a grid:

1. We'll start by first defining the mapping between different cell states and their color codes to be used in the Gridworld environment:

```
EMPTY = BLACK = 0
WALL = GRAY = 1
AGENT = BLUE = 2
MINE = RED = 3
GOAL = GREEN = 4
SUCCESS = PINK = 5
```

2. Next, generate a color map using RGB intensity values:

```
COLOR_MAP = {
    BLACK: [0.0, 0.0, 0.0],
    GRAY: [0.5, 0.5, 0.5],
    BLUE: [0.0, 0.0, 1.0],
    RED: [1.0, 0.0, 0.0],
    GREEN: [0.0, 1.0, 0.0],
    PINK: [1.0, 0.0, 1.0],
}
```

3. Let's now define the action mapping:

```
NOOP = 0
DOWN = 1
UP = 2
LEFT = 3
RIGHT = 4
```

4. Let's then create a GridworldEnv class with an __init__ function to define necessary class variables, including the observation and action space:

```
class GridworldEnv():
    def __init__(self):
```

We will implement __init__() in the following steps.

5. In this step, let's define the layout of the Gridworld environment using the grid cell state mapping:

```
self.grid_layout = """
1 1 1 1 1 1 1 1
1 2 0 0 0 0 0 1
1 0 1 1 1 0 0 1
1 0 1 0 1 0 0 1
1 0 1 4 1 0 0 1
1 0 3 0 0 0 0 1
1 0 0 0 0 0 0 1
1 1 1 1 1 1 1 1
"""
```

In the preceding layout, 0 corresponds to the empty cells, 1 corresponds to walls, 2 corresponds to the agent's starting location, 3 corresponds to the location of the mine/bomb/obstacle, and 4 corresponds to the goal location based on the mapping we defined in step 1.

6. Now, we are ready to define the observation space for the Gridworld RL environment:

```
self.initial_grid_state = np.fromstring(
                 self.grid_layout, dtype=int, sep=" ")
self.initial_grid_state = \
                 self.initial_grid_state.reshape(8, 8)
self.grid_state = copy.deepcopy(
                 self.initial_grid_state)
self.observation_space = gym.spaces.Box(
        low=0, high=6, shape=self.grid_state.shape
)
self.img_shape = [256, 256, 3]
self.metadata = {"render.modes": ["human"]}
```

7. Let's define the action space and the mapping between the actions and the movement of the agent in the grid:

```
self.action_space = gym.spaces.Discrete(5)
self.actions = [NOOP, UP, DOWN, LEFT, RIGHT]
self.action_pos_dict = {
```

```
    NOOP: [0, 0],
    UP: [-1, 0],
    DOWN: [1, 0],
    LEFT: [0, -1],
    RIGHT: [0, 1],
}
```

8. Let's now wrap up the __init__ function by initializing the agent's start and goal states using the get_state() method (which we will implement in the next step):

```
(self.agent_start_state, self.agent_goal_state,) = \
                                      self.get_state()
```

9. Now we need to implement the get_state() method that returns the start and goal state for the Gridworld environment:

```
def get_state(self):
        start_state = np.where(self.grid_state == AGENT)
        goal_state = np.where(self.grid_state == GOAL)

        start_or_goal_not_found = not (start_state[0] \
                                      and goal_state[0])
        if start_or_goal_not_found:
            sys.exit(
                "Start and/or Goal state not present in
                 the Gridworld. "
                "Check the Grid layout"
            )
        start_state = (start_state[0][0],
                       start_state[1][0])
        goal_state = (goal_state[0][0], goal_state[1][0])

        return start_state, goal_state
```

10. In this step, we will be implementing the `step(action)` method to execute the action and retrieve the next state/observation, the associated reward, and whether the episode ended:

```python
def step(self, action):
    """return next observation, reward, done, info"""
    action = int(action)
    info = {"success": True}
    done = False
    reward = 0.0
    next_obs = (
        self.agent_state[0] + \
            self.action_pos_dict[action][0],
        self.agent_state[1] + \
            self.action_pos_dict[action][1],
    )
```

11. Next, let's specify the rewards and finally, return `grid_state`, `reward`, `done`, and `info`:

```python
    # Determine the reward
    if action == NOOP:
        return self.grid_state, reward, False, info
    next_state_valid = (
        next_obs[0] < 0 or next_obs[0] >= \
                        self.grid_state.shape[0]
    ) or (next_obs[1] < 0 or next_obs[1] >= \
                        self.grid_state.shape[1])
    if next_state_valid:
        info["success"] = False
        return self.grid_state, reward, False, info

    next_state = self.grid_state[next_obs[0],
                                 next_obs[1]]
    if next_state == EMPTY:
        self.grid_state[next_obs[0],
                        next_obs[1]] = AGENT
    elif next_state == WALL:
```

```
        info["success"] = False
        reward = -0.1
        return self.grid_state, reward, False, info
    elif next_state == GOAL:
        done = True
        reward = 1
    elif next_state == MINE:
        done = True
        reward = -1            # self._render("human")
    self.grid_state[self.agent_state[0],
                    self.agent_state[1]] = EMPTY
    self.agent_state = copy.deepcopy(next_obs)
    return self.grid_state, reward, done, info
```

12. Up next is the `reset()` method, which resets the Gridworld environment when an episode completes (or if a request to reset the environment is made):

```
def reset(self):
    self.grid_state = copy.deepcopy(
                        self.initial_grid_state)
    (self.agent_state, self.agent_goal_state,) = \
                        self.get_state()
    return self.grid_state
```

13. To visualize the state of the Gridworld environment in a human-friendly manner, let's implement a render function that will convert the `grid_layout` that we defined in step 5 to an image and display it. With that, the Gridworld environment implementation will be complete!

```
def gridarray_to_image(self, img_shape=None):
    if img_shape is None:
        img_shape = self.img_shape
    observation = np.random.randn(*img_shape) * 0.0
    scale_x = int(observation.shape[0] / self.grid_\
                    state.shape[0])
    scale_y = int(observation.shape[1] / self.grid_\
                    state.shape[1])
    for i in range(self.grid_state.shape[0]):
```

```
        for j in range(self.grid_state.shape[1]):
            for k in range(3):  # 3-channel RGB image
                pixel_value = \
                    COLOR_MAP[self.grid_state[i, j]][k]
                observation[
                    i * scale_x : (i + 1) * scale_x,
                    j * scale_y : (j + 1) * scale_y,
                    k,
                ] = pixel_value
    return (255 * observation).astype(np.uint8)


def render(self, mode="human", close=False):
    if close:
        if self.viewer is not None:
            self.viewer.close()
            self.viewer = None
        return

    img = self.gridarray_to_image()
    if mode == "rgb_array":
        return img
    elif mode == "human":
        from gym.envs.classic_control import \
            rendering

        if self.viewer is None:
            self.viewer = \
                    rendering.SimpleImageViewer()
        self.viewer.imshow(img)
```

14. To test whether the environment is working as expected, let's add a __main__ function that gets executed if the environment script is run directly:

```
if __name__ == "__main__":
    env = GridworldEnv()
    obs = env.reset()
    # Sample a random action from the action space
```

```
action = env.action_space.sample()
next_obs, reward, done, info = env.step(action)
print(f"reward:{reward} done:{done} info:{info}")
env.render()
env.close()
```

15. All set! The Gridworld environment is ready and we can quickly test it by running the script (`python envs/gridworld.py`). An output such as the following will be displayed:

```
reward:0.0 done:False info:{'success': True}
```

The following rendering of the Gridworld environment will also be displayed:

Figure 1.2 – The Gridworld

Let's now see how it works!

How it works...

The `grid_layout` defined in step 5 in the *How to do it...* section represents the state of the learning environment. The Gridworld environment defines the observation space, action spaces, and the rewarding mechanism to implement a **Markov Decision Process (MDP)**. We sample a valid action from the action space of the environment and step the environment with the chosen action, which results in the new observation, reward, and a done status Boolean (representing if the episode has finished) as the response from the Gridworld environment. The `env.render()` method converts the environment's internal grid representation to an image and displays it for visual understanding.

Implementing neural network-based RL policies for discrete action spaces and decision-making problems

Many environments (both simulated and real) for RL requires the RL agent to choose an action from a list of actions or, in other words, take discrete actions. While simple linear functions can be used to represent policies for such agents, they are often not scalable to complex problems. A non-linear function approximator such as a (deep) neural network can approximate arbitrary functions, even those required to solve complex problems.

The neural network-based policy network is a crucial building block for advanced RL and **Deep RL** and will be applicable to general, discrete decision-making problems.

By the end of this recipe, you will have an agent with a neural network-based policy implemented in **TensorFlow 2.x** that can take actions in the **Gridworld** environment and (with little or no modifications) in any discrete-action space environment.

Getting ready

Activate the `tf2rl-cookbook` Python virtual environment and run the following to install and import the packages:

```
pip install --upgrade numpy tensorflow tensorflow_probability
seaborn
```

```
import seaborn as sns
```

```
import tensorflow as tf
```

```
from tensorflow import keras
```

```
from tensorflow.keras import layers
```

```
import tensorflow_probability as tfp
```

Let's get started.

How to do it...

We will look at policy distribution types that can be used by agents in environments with discrete action spaces:

1. Let's begin by creating a binary policy distribution in TensorFlow 2.x using the `tensorflow_probability` library:

```
binary_policy = tfp.distributions.Bernoulli(probs=0.5)
for i in range(5):
    action = binary_policy.sample(1)
    print("Action:", action)
```

The preceding code should print something like the following:

```
Action: tf.Tensor([0], shape=(1,), dtype=int32)
Action: tf.Tensor([1], shape=(1,), dtype=int32)
Action: tf.Tensor([0], shape=(1,), dtype=int32)
Action: tf.Tensor([1], shape=(1,), dtype=int32)
Action: tf.Tensor([1], shape=(1,), dtype=int32)
```

> **Important note**
>
> The values of the action that you get will differ from what is shown here because they will be sampled from the Bernoulli distribution, which is not a deterministic process.

2. Let's quickly visualize the binary policy distribution:

```
# Sample 500 actions from the binary policy distribution
sample_actions = binary_policy.sample(500)
sns.distplot(sample_actions)
```

The preceding code will generate a distribution plot as shown here:

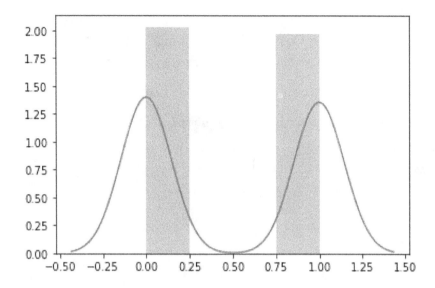

Figure 1.3 – A distribution plot of the binary policy

3. In this step, we will be implementing a discrete policy distribution. A categorical distribution over a single discrete variable with k finite categories is referred to as a **multinoulli** distribution. The generalization of the multinoulli distribution to multiple trials is the multinomial distribution that we will be using to represent discrete policy distributions:

```
action_dim = 4  # Dimension of the discrete action space
action_probabilities = [0.25, 0.25, 0.25, 0.25]
discrete_policy = tfp.distributions.
Multinomial(probs=action_probabilities, total_count=1)
for i in range(5):
    action = discrete_policy.sample(1)
    print(action)
```

The preceding code should print something along the lines of the following:

> **Important note**
> The values of the action that you get will differ from what is shown here because they will be sampled from the multinomial distribution, which is not a deterministic process.

```
tf.Tensor([[0. 0. 0. 1.]], shape=(1, 4), dtype=float32)
tf.Tensor([[0. 0. 1. 0.]], shape=(1, 4), dtype=float32)
```

```
tf.Tensor([[0. 0. 1. 0.]], shape=(1, 4), dtype=float32)
tf.Tensor([[1. 0. 0. 0.]], shape=(1, 4), dtype=float32)
tf.Tensor([[0. 1. 0. 0.]], shape=(1, 4), dtype=float32)
```

4. Next, we visualize the discrete probability distribution:

```
sns.distplot(discrete_policy.sample(1))
```

The preceding code will generate a distribution plot, like the one shown here for `discrete_policy`:

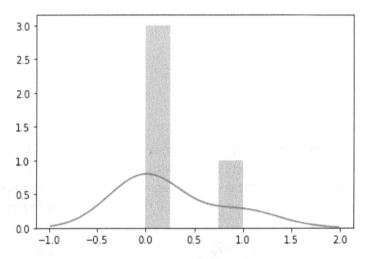

Figure 1.4 – A distribution plot of the discrete policy

5. Then, calculate the entropy of a discrete policy:

```
def entropy(action_probs):
    return -tf.reduce_sum(action_probs * \
                    tf.math.log(action_probs), axis=-1)
action_probabilities = [0.25, 0.25, 0.25, 0.25]
print(entropy(action_probabilities))
```

6. Also, implement a discrete policy class:

```
class DiscretePolicy(object):
    def __init__(self, num_actions):
        self.action_dim = num_actions
    def sample(self, actino_logits):
        self.distribution = tfp.distributions.
```

```
Multinomial(logits=action_logits, total_count=1)
        return self.distribution.sample(1)
    def get_action(self, action_logits):
        action = self.sample(action_logits)
        return np.where(action)[-1]
        # Return the action index
    def entropy(self, action_probabilities):
        return - tf.reduce_sum(action_probabilities *
tf.math.log(action_probabilities), axis=-1)
```

7. Now we implement a helper method to evaluate the agent in a given environment:

```
def evaluate(agent, env, render=True):
    obs, episode_reward, done, step_num = env.reset(),
                                    0.0, False, 0
    while not done:
        action = agent.get_action(obs)
        obs, reward, done, info = env.step(action)
        episode_reward += reward
        step_num += 1
        if render:
            env.render()
    return step_num, episode_reward, done, info
```

8. Let's now implement a neural network Brain class using TensorFlow 2.x:

```
class Brain(keras.Model):
    def __init__(self, action_dim=5,
                    input_shape=(1, 8 * 8)):
        """Initialize the Agent's Brain model

        Args:
            action_dim (int): Number of actions
        """
        super(Brain, self).__init__()
        self.dense1 = layers.Dense(32, input_shape=\
                        input_shape, activation="relu")
        self.logits = layers.Dense(action_dim)
```

```python
    def call(self, inputs):
        x = tf.convert_to_tensor(inputs)
        if len(x.shape) >= 2 and x.shape[0] != 1:
            x = tf.reshape(x, (1, -1))
        return self.logits(self.dense1(x))


    def process(self, observations):
    # Process batch observations using `call(inputs)` behind-
    the-scenes
        action_logits = \
                    self.predict_on_batch(observations)
        return action_logits
```

9. Let's now implement a simple agent class that uses a `DiscretePolicy` object to act in discrete environments:

```python
class Agent(object):
    def __init__(self, action_dim=5,
                    input_dim=(1, 8 * 8)):
        self.brain = Brain(action_dim, input_dim)
        self.policy = DiscretePolicy(action_dim)
    def get_action(self, obs):
        action_logits = self.brain.process(obs)
        action = self.policy.get_action(
                        np.squeeze(action_logits, 0))
        return action
```

10. Let's now test the agent in `GridworldEnv`:

```python
from envs.gridworld import GridworldEnv
env = GridworldEnv()
agent = Agent(env.action_space.n,
            env.observation_space.shape)
steps, reward, done, info = evaluate(agent, env)
print(f"steps:{steps} reward:{reward} done:{done}
info:{info}")
env.close()
```

This shows how to implement the policy. We will see how this works in the following section.

How it works...

One of the central components of an RL agent is the policy function that maps between observations and actions. Formally, a policy is a distribution over actions that prescribes the probabilities of choosing an action given an observation.

In environments where the agent can take at most two different actions, for example, in a binary action space, we can represent the policy using a **Bernoulli distribution**, where the probability of taking action 0 is given by $p(x = 1) = \phi$, and the probability of taking action 1 is given by $p(x = 0) = 1 - \phi$, which gives rise to the following probability distribution:

$$p(x = x) = \phi^x(1 - \phi)^{(1-x)}$$

A discrete probability distribution can be used to represent an RL agent's policy when the agent can take one of k possible actions in an environment.

In a general sense, such distributions can be used to describe the possible results of a random variable that can take one of k possible categories and is therefore also called a **categorical distribution**. This is a generalization of the Bernoulli distribution to k-way events and is therefore a multinoulli distribution.

Implementing neural network-based RL policies for continuous action spaces and continuous-control problems

Reinforcement learning has been used to achieve the state of the art in many control problems, not only in games as varied as Atari, Go, Chess, Shogi, and StarCraft, but also in real-world deployments, such as HVAC control systems.

In environments where the action space is continuous, meaning that the actions are real-valued, a real-valued, continuous policy distribution is necessary. A continuous probability distribution can be used to represent an RL agent's policy when the action space of the environment contains real numbers. In a general sense, such distributions can be used to describe the possible results of a random variable when the random variable can take any (real) value.

Once the recipe is complete, you will have a complete script to control a car in two dimensions to drive up a hill using the MountainCarContinuous environment with a continuous action space. A screenshot from the MountainCarContinuous environment is shown here:

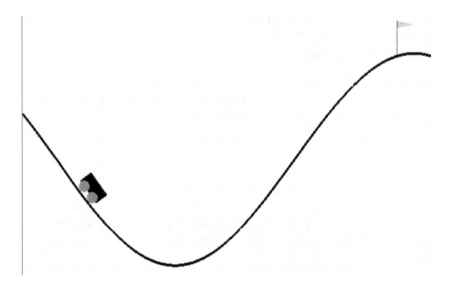

Figure 1.5 – A screenshot of the MountainCarContinuous environment

Getting ready

Activate the tf2rl-cookbook Conda Python environment and run the following command to install and import the necessary Python packages for this recipe:

```
pip install --upgrade tensorflow_probability
import tensorflow_probability as tfp
import seaborn as sns
```

Let's get started.

How to do it...

We will begin by creating continuous policy distributions using **TensorFlow 2.x** and the `tensorflow_probability` library and build upon the necessary action sampling methods to generate action for a given continuous space of an RL environment:

1. We create a continuous policy distribution in TensorFlow 2.x using the `tensorflow_probability` library. We will use a Gaussian/normal distribution to create a policy distribution over continuous values:

    ```
    sample_actions = continuous_policy.sample(500)
    sns.distplot(sample_actions)
    ```

2. Next, we visualize a continuous policy distribution:

    ```
    sample_actions = continuous_policy.sample(500)
    sns.distplot(sample_actions)
    ```

    The preceding code will generate a distribution plot of the continuous policy, like the plot shown here:

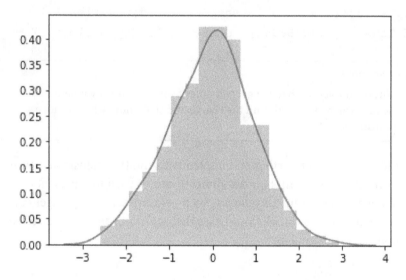

Figure 1.6 – A distribution plot of the continuous policy

3. Let's now implement a continuous policy distribution using a Gaussian/normal distribution:

    ```
    mu = 0.0  # Mean = 0.0
    sigma = 1.0  # Std deviation = 1.0
    ```

```
continuous_policy = tfp.distributions.Normal(loc=mu,
                                              scale=sigma)
# action = continuous_policy.sample(10)
for i in range(10):
    action = continuous_policy.sample(1)
    print(action)
```

The preceding code should print something similar to what is shown in the following code block:

```
tf.Tensor([-0.2527136], shape=(1,), dtype=float32)
tf.Tensor([1.3262751], shape=(1,), dtype=float32)
tf.Tensor([0.81889665], shape=(1,), dtype=float32)
tf.Tensor([1.754675], shape=(1,), dtype=float32)
tf.Tensor([0.30025303], shape=(1,), dtype=float32)
tf.Tensor([-0.61728036], shape=(1,), dtype=float32)
tf.Tensor([0.40142158], shape=(1,), dtype=float32)
tf.Tensor([1.3219402], shape=(1,), dtype=float32)
tf.Tensor([0.8791297], shape=(1,), dtype=float32)
tf.Tensor([0.30356944], shape=(1,), dtype=float32)
```

> **Important note**
> The values of the action that you get will differ from what is shown here because they will be sampled from the Gaussian distribution, which is not a deterministic process.

4. Let's now move one step further and implement a multi-dimensional continuous policy. A **multivariate Gaussian distribution** can be used to represent multi-dimensional continuous policies. Such polices are useful for agents when acting in environments with action spaces that are multi-dimensional, as well as continuous and real-valued:

```
mu = [0.0, 0.0]
covariance_diag = [3.0, 3.0]
continuous_multidim_policy = tfp.distributions.
MultivariateNormalDiag(loc=mu, scale_diag=covariance_
diag)
# action = continuous_multidim_policy.sample(10)
for i in range(10):
```

```
action = continuous_multidim_policy.sample(1)
print(action)
```

The preceding code should print something similar to what follows:

> **Important note**
>
> The values of the action that you get will differ from what is shown here because they will be sampled from the multivariate Gaussian/normal distribution, which is not a deterministic process).

```
tf.Tensor([[ 1.7003113 -2.5801306]], shape=(1, 2),
dtype=float32)
```

```
tf.Tensor([[ 2.744986  -0.5607129]], shape=(1, 2),
dtype=float32)
```

```
tf.Tensor([[ 6.696332  -3.3528223]], shape=(1, 2),
dtype=float32)
```

```
tf.Tensor([[ 1.2496299 -8.301748 ]], shape=(1, 2),
dtype=float32)
```

```
tf.Tensor([[2.0009246 3.557394 ]], shape=(1, 2),
dtype=float32)
```

```
tf.Tensor([[-4.491785  -1.0101566]], shape=(1, 2),
dtype=float32)
```

```
tf.Tensor([[ 3.0810184 -0.9008362]], shape=(1, 2),
dtype=float32)
```

```
tf.Tensor([[1.4185237 2.2145705]], shape=(1, 2),
dtype=float32)
```

```
tf.Tensor([[-1.9961193 -2.1251974]], shape=(1, 2),
dtype=float32)
```

```
tf.Tensor([[-1.2200387 -4.3516426]], shape=(1, 2),
dtype=float32)
```

5. Before moving on, let's visualize the multi-dimensional continuous policy:

```
sample_actions = continuous_multidim_policy.sample(500)
sns.jointplot(sample_actions[:, 0], sample_actions[:, 1],
kind='scatter')
```

The preceding code will generate a joint distribution plot similar to the plot shown here:

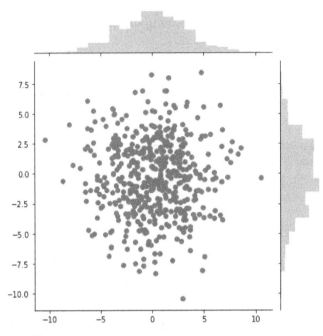

Figure 1.7 – Joint distribution plot of a multi-dimensional continuous policy

6. Now, we are ready to implement the continuous policy class:

```
class ContinuousPolicy(object):
    def __init__(self, action_dim):
        self.action_dim = action_dim
    def sample(self, mu, var):
        self.distribution = \
            tfp.distributions.Normal(loc=mu, scale=sigma)
        return self.distribution.sample(1)
    def get_action(self, mu, var):
        action = self.sample(mu, var)
        return action
```

7. As a next step, let's implement a multi-dimensional continuous policy class:

```
import tensorflow_probability as tfp
import numpy as np
class ContinuousMultiDimensionalPolicy(object):
```

```python
    def __init__(self, num_actions):
        self.action_dim = num_actions
    def sample(self, mu, covariance_diag):
        self.distribution = tfp.distributions.\
                        MultivariateNormalDiag(loc=mu,
                        scale_diag=covariance_diag)
        return self.distribution.sample(1)
    def get_action(self, mu, covariance_diag):
        action = self.sample(mu, covariance_diag)
        return action
```

8. Let's now implement a function to evaluate an agent in an environment with a continuous action space to assess episodic performance:

```python
def evaluate(agent, env, render=True):
    obs, episode_reward, done, step_num = env.reset(),
                        0.0, False, 0
    while not done:
        action = agent.get_action(obs)
        obs, reward, done, info = env.step(action)
        episode_reward += reward
        step_num += 1
        if render:
            env.render()
    return step_num, episode_reward, done, info
```

9. We are now ready to test the agent in a continuous action environment:

```python
from neural_agent import Brain
import gym
env = gym.make("MountainCarContinuous-v0") Implementing a
Neural-network Brain class using TensorFlow 2.x.


        class Brain(keras.Model):
    def __init__(self, action_dim=5,
                input_shape=(1, 8 * 8)):
        """Initialize the Agent's Brain model
```

```
    Args:
        action_dim (int): Number of actions
    """
    super(Brain, self).__init__()
    self.dense1 = layers.Dense(32,
            input_shape=input_shape, activation="relu")
    self.logits = layers.Dense(action_dim)

def call(self, inputs):
    x = tf.convert_to_tensor(inputs)
    if len(x.shape) >= 2 and x.shape[0] != 1:
        x = tf.reshape(x, (1, -1))
    return self.logits(self.dense1(x))

def process(self, observations):
    # Process batch observations using `call(inputs)`
    # behind-the-scenes
    action_logits = \
        self.predict_on_batch(observations)
    return action_logits
```

10. Let's implement a simple agent class that utilizes the ContinuousPolicy object to act in continuous action space environments:

```
class Agent(object):
    def __init__(self, action_dim=5,
                    input_dim=(1, 8 * 8)):
        self.brain = Brain(action_dim, input_dim)
        self.policy = ContinuousPolicy(action_dim)
    def get_action(self, obs):
        action_logits = self.brain.process(obs)
        action = self.policy.get_action(*np.\
                        squeeze(action_logits, 0))
        return action
```

11. As a final step, we will test the performance of the agent in a continuous action space environment:

```
from neural_agent import Brain
import gym
env = gym.make("MountainCarContinuous-v0")

action_dim = 2 * env.action_space.shape[0]
    # 2 values (mu & sigma) for one action dim
agent = Agent(action_dim, env.observation_space.shape)
steps, reward, done, info = evaluate(agent, env)
print(f"steps:{steps} reward:{reward} done:{done}
info:{info}")
env.close()
```

The preceding script will call the MountainCarContinuous environment, render it to the screen, and show how the agent is performing in this continuous action space environment:

Figure 1.8 – A screenshot of the agent in the MountainCarContinuous-v0 environment

Next, let's explore how it works.

How it works...

We implemented a continuous-valued policy for RL agents using a **Gaussian distribution**. Gaussian distribution, which is also known as **normal distribution**, is the most widely used distribution for real numbers. It is represented using two parameters, μ and σ. We generated continuous-valued actions from such a policy by sampling from the distribution, based on the probability density that is given by the following equation:

$$N(x; \mu, \sigma^2) = \sqrt{\frac{1}{2\pi\sigma^2}} \exp(-\frac{1}{(2\sigma^2)(x-\mu)^2})$$

The **multivariate normal distribution** extends the normal distribution to multiple variables. We used this distribution to generate multi-dimensional continuous policies.

Working with OpenAI Gym for RL training environments

This recipe provides a quick run-through for getting up and running with OpenAI Gym environments. The Gym environment and the interface provide a platform for training RL agents and is the most widely used and accepted RL environment interface.

Getting ready

We will be needing the full installation of OpenAI Gym to be able to use the available environments. Please follow the Gym installation steps listed at https://github. com/openai/gym#id5.

As a minimum, you should execute the following command:

```
pip install gym[atari]
```

How to do it...

Let's start by picking an environment and exploring the Gym interface. You may already be familiar with the basic function calls to create a Gym environment from the previous recipes.

Your steps should be formatted like so:

1. Let's first explore the list of environments in Gym:

    ```python
    #!/usr/bin/env python
    from gym import envs
    env_names = [spec.id for spec in envs.registry.all()]
    for name in sorted(env_names):
        print(name)
    ```

2. This script will print the names of all the environments available through your Gym installation, sorted alphabetically. You can run this script using the following command to see the names of the environments that are installed and available in your system. You should see a long list of environments listed. The first few are shown in the following screenshot for your reference:

Figure 1.9 – List of environments available using the openai-gym package

Let's now see how we can run one of the Gym environments.

3. The following script will let you explore any of the available Gym environments:

```python
#!/usr/bin/env python


import gym
import sys


def run_gym_env(argv):
    env = gym.make(argv[1])  # Name of the environment
                             # supplied as 1st argument
    env.reset()
```

```python
        for _ in range(int(argv[2])):
            env.render()
            env.step(env.action_space.sample())
        env.close()


if __name__ == "__main__":
    run_gym_env(sys.argv)
```

4. You can save the preceding script to `run_gym_env.py` and run the script like this:

```
(tf2rl-cookbook) praveen@g5: ~/tf2rl-cookbook/ch1/
src$python run_gym_env.py Alien-v4 1000
```

The script will render the `Alien-v4` environment, which should look like the following screenshot:

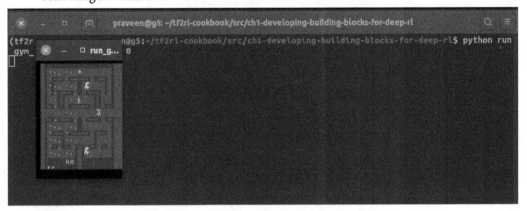

Figure 1.10 – Sample output of run_gym_env.py with Alien-v4 1000 as the arguments

> **Tip**
> You can change `Alien-v4` to any of the available Gym environments listed in the previous step.

How it works...

A summary of how the Gym environments work is presented in the following table:

Returned Value	Type	Description
next_observation	Object	Observation returned by the environment. The object could be the RGB pixel data from the screen/camera, RAM contents, join angles and join velocities of a robot, and so on, depending on the environment.
Reward	Float	Reward for the previous action that was sent to the environment. The range of the Float value varies with each environment, but irrespective of the environment, a higher reward is always better and the goal of the agent should be to maximize the total reward.
Done	Boolean	Indicates whether the environment is going to be reset in the next step. When the Boolean value is true, it most likely means that the episode has ended (due to loss of life of the agent, timeout, or some other episode termination criteria).
Info	Dict	Some additional information that can optionally be sent out by an environment as a dictionary of arbitrary key-value pairs. The agent we develop should not rely on any of the information in this dictionary for taking action. It may be used (if available) for debugging purposes.

Table 1.1 – Summary of the Gym environment interface

See also

You can find more information on OpenAI Gym here: `http://gym.openai.com/`.

Building a neural agent

This recipe will guide you through the steps to build a complete agent and the agent-environment interaction loop, which is the main building block for any RL application. When you complete the recipe, you will have an executable script where a simple agent tries to act in a Gridworld environment. A glimpse of what the agent you build will likely be doing is shown in the following screenshot:

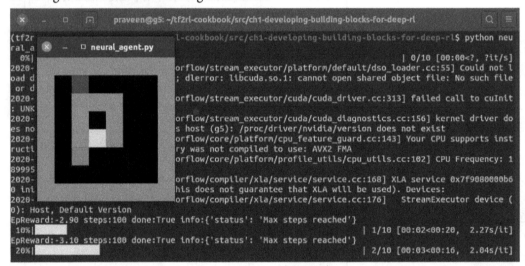

Figure 1.11 – Screenshot of output from the neural_agent.py script

Getting ready

Let's get started by activating the tf2rl-cookbook Conda Python environment and running the following code to install and import the necessary Python modules:

```
pip install tensorflow gym tqdm  # Run this line in a terminal

import tensorflow as tf

from tensorflow import keras

from tensorflow.keras import layers

import gym

import envs

from tqdm import tqdm
```

How to do it...

We will start by implementing a Brain class powered by a neural network implemented using TensorFlow 2.x:

1. Let's first initialize a neural brain model using TensorFlow 2.x and the Keras functional API:

```python
class Brain(keras.Model):
    def __init__(self, action_dim=5,
                 input_shape=(1, 8 * 8)):
        """Initialize the Agent's Brain model

        Args:
            action_dim (int): Number of actions
        """
        super(Brain, self).__init__()
        self.dense1 = layers.Dense(32, input_shape= \
                        input_shape, activation="relu")
        self.logits = layers.Dense(action_dim)
```

2. Next, we implement the Brain class's call (...) method:

```python
def call(self, inputs):
    x = tf.convert_to_tensor(inputs)
    if len(x.shape) >= 2 and x.shape[0] != 1:
        x = tf.reshape(x, (1, -1))
    return self.logits(self.dense1(x))
```

3. Now we need to implement the Brain class's process() method to conveniently perform predictions on a batch of inputs/observations:

```python
def process(self, observations):
    # Process batch observations using `call(inputs)`
    # behind-the-scenes
    action_logits = \
        self.predict_on_batch(observations)
    return action_logits
```

4. Let's now implement the init function of the agent class:

```
class Agent(object):
    def __init__(self, action_dim=5,
                    input_shape=(1, 8 * 8)):
        """Agent with a neural-network brain powered
            policy

        Args:
            brain (keras.Model): Neural Network based
        model
        """
        self.brain = Brain(action_dim, input_shape)
        self.policy = self.policy_mlp
```

5. Now let's define a simple policy function for the agent:

```
def policy_mlp(self, observations):
    observations = observations.reshape(1, -1)
    # action_logits = self.brain(observations)
    action_logits = self.brain.process(observations)
    action = tf.random.categorical(tf.math.\
                log(action_logits), num_samples=1)
    return tf.squeeze(action, axis=1)
```

6. After that, let's implement a convenient `get_action` method for the agent:

```
def get_action(self, observations):
    return self.policy(observations)
```

7. Let's now create a placeholder function for `learn()` that will be implemented as part of RL algorithm implementation in future recipes:

```
def learn(self, samples):
    raise NotImplementedError
```

This completes our basic agent implementation with the necessary ingredients!

8. Let's now evaluate the agent in a given environment for one episode:

```python
def evaluate(agent, env, render=True):
    obs, episode_reward, done, step_num = env.reset(),
                                          0.0, False, 0
    while not done:
        action = agent.get_action(obs)
        obs, reward, done, info = env.step(action)
        episode_reward += reward
        step_num += 1
        if render:
            env.render()
    return step_num, episode_reward, done, info
```

9. Finally, let's implement the main function:

```python
if __name__ == "__main__":
    env = gym.make("Gridworld-v0")
    agent = Agent(env.action_space.n,
                  env.observation_space.shape)
    for episode in tqdm(range(10)):
        steps, episode_reward, done, info = \
                                    evaluate(agent, env)
        print(f"EpReward:{episode_reward:.2f}\
            steps:{steps} done:{done} info:{info}")
    env.close()
```

10. Execute the script as follows:

```
python neural_agent.py
```

You should see the Gridworld environment GUI pop up. This will show you what the agent is doing in the environment, and it will look like the following screenshot:

Figure 1.12 – A screenshot of the neural agent acting in the Gridworld environment

This provides a simple, yet complete, recipe to build an agent and the agent-environment interaction loop. All that is left is to add the RL algorithm of your choice to the `learn()` method and the agent will start acting intelligently!

How it works...

This recipe puts together the necessary ingredients to build a complete agent-environment system. The `Brain` class implements the neural network that serves as the processing unit of the agent, and the agent class utilizes the `Brain` class and a simple policy that chooses an action based on the output of the brain after processing the observations received from the environment.

We implemented the `Brain` class as a subclass of the `keras.Model` class, which allows us to define a custom neural network-based model for the agent's brain. The `__init__` method initializes the `Brain` model and defines the necessary layers using the **TensorFlow 2.x Keras functional API**. In this `Brain` model, we are creating two **dense** (also known as **fully-connected**) layers to build our starter neural network. In addition to the `__init__` method, the `call(...)` method is also a mandatory method that needs to be implemented by child classes inheriting from the `keras.Model` class. The `call(...)` method first converts the inputs to a TensorFlow 2.x tensor and then flattens the inputs to be of the shape `1 x total_number_of_elements` in the input tensor. For example, if the input data has a shape of 8 x 8 (8 rows and 8 columns), the data is first converted to a tensor and the shape is flattened to 1 x 8 * 8 = 1 x 64. The flattened inputs are then processed by the dense1 layer, which contains 32 neurons and a ReLU activation function. Finally, the logits layer processes the output from the previous layer and produces n number of outputs corresponding to the action dimension (n).

The `predict_on_batch(...)` method performs predictions on the batch of inputs given as the argument. This function (unlike the `predict()` function of **Keras**) assumes that the inputs (observations) provided as the argument are exactly one batch of inputs and thus feeds the batch to the network without any further splitting of the input data.

We then implemented the `Agent` class and, in the agent initialization function, we created an object instance of the Brain class by defining the following:

```
self.brain = Brain(action_dim, input_shape)
```

Here, `input_shape` is the shape of the input that is expected to be processed by the brain, and `action_dim` is the shape of the output expected from the brain. The agent's policy is defined to be a custom **Multi-Layer Perceptron (MLP)**-based policy based on the brain's neural network architecture. Note that we can reuse `DiscretePolicy` from the previous recipe to initialize the agent's policy as well.

The agent's policy function, `policy_mlp`, flattens the input observations and sends it for processing by the agent's brain to receive the `action_logits`, which are the unnormalized probabilities for the actions. The final action to be taken is obtained by using TensorFlow 2.x's `categorical` method from the random module, which samples a valid action from the given `action_logits` (unnormalized probabilities).

> **Important note**
> If all of the observations supplied to the `predict_on_batch` function cannot be accommodated in the given amount of GPU memory or RAM (CPU), the operation can cause a GPU **Out Of Memory (OOM)** error.

The main function that gets launched – if the `neural_agent.py` script is run directly – creates an instance of the Gridworld-v0 environment, initializes an agent using the action and observation space of this environment, and starts evaluating the agent for 10 episodes.

Building a neural evolutionary agent

Evolutionary methods are based on black-box optimization and are also known as gradient-free methods since no gradient computation is involved. This recipe will walk you through the steps for implementing a simple, approximate cross-entropy-based neural evolutionary agent using **TensorFlow 2.x.**

Getting ready

Activate the `tf2rl-cookbook` Python environment and import the following packages necessary to run this recipe:

```python
from collections import namedtuple

import gym
import matplotlib.pyplot as plt
import numpy as np
import tensorflow as tf
from tensorflow import keras
from tensorflow.keras import layers
from tqdm import tqdm

import envs
```

With the packages installed, we are ready to begin.

How to do it...

Let's put together all that we have learned in this chapter to build a neural agent that improves its policy to navigate the Gridworld environment using an evolutionary process:

1. Let's start by importing the basic neural agent and the Brain class from `neural_agent.py`:

```
from neural_agent import Agent, Brain
from envs.gridworld import GridworldEnv
```

2. Next, let's implement a method to roll out the agent in a given environment for one episode and return `obs_batch`, `actions_batch`, and `episode_reward`:

```
def rollout(agent, env, render=False):
    obs, episode_reward, done, step_num = env.reset(),
                                          0.0, False, 0
    observations, actions = [], []
    episode_reward = 0.0
    while not done:
        action = agent.get_action(obs)
        next_obs, reward, done, info = env.step(action)
        # Save experience
        observations.append(np.array(obs).reshape(1, -1))
        # Convert to numpy & reshape (8, 8) to (1, 64)
        actions.append(action)
        episode_reward += reward

        obs = next_obs
        step_num += 1
        if render:
            env.render()
    env.close()
    return observations, actions, episode_reward
```

3. Let's now test the trajectory rollout method:

```
env = GridworldEnv()
# input_shape = (env.observation_space.shape[0] * \
                env.observation_space.shape[1], )
brain = Brain(env.action_space.n)
agent = Agent(brain)
obs_batch, actions_batch, episode_reward = rollout(agent,
                                                   env)
```

4. Now, it's time for us to verify that the experience data generated using the rollouts is coherent:

```
assert len(obs_batch) == len(actions_batch)
```

5. Let's now roll out multiple complete trajectories to collect experience data:

```
# Trajectory: (obs_batch, actions_batch, episode_reward)
# Rollout 100 episodes; Maximum possible steps = 100 *
100 = 10e4
trajectories = [rollout(agent, env, render=True) \
               for _ in tqdm(range(100))]
```

6. We can then visualize the reward distribution from a sample of experience data. Let's also plot a red vertical line at the 50th percentile of the episode reward values in the collected experience data:

```
from tqdm.auto import tqdm
import matplotlib.pyplot as plt
%matplotlib inline

sample_ep_rewards = [rollout(agent, env)[-1] for _ in \
                    tqdm(range(100))]

plt.hist(sample_ep_rewards, bins=10, histtype="bar");
```

Running this code will generate a plot like the one shown in the following diagram:

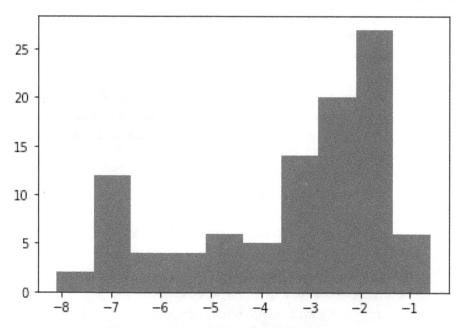

Figure 1.13 – Histogram plot of the episode reward values

7. Let's now create a container for storing trajectories:

```python
from collections import namedtuple
Trajectory = namedtuple("Trajectory", ["obs", "actions",
                                        "reward"])
# Example for understanding the operations:
print(Trajectory(*(1, 2, 3)))
# Explanation: `*` unpacks the tuples into individual
# values
Trajectory(*(1, 2, 3)) == Trajectory(1, 2, 3)
# The rollout(...) function returns a tuple of 3 values:
# (obs, actions, rewards)
# The Trajectory namedtuple can be used to collect
# and store mini batch of experience to train the neuro
# evolution agent
trajectories = [Trajectory(*rollout(agent, env)) \
                for _ in range(2)]
```

8. Now it's time to choose elite experiences for the evolution process:

```python
def gather_elite_xp(trajectories, elitism_criterion):
    """Gather elite trajectories from the batch of
        trajectories
    Args:
        batch_trajectories (List): List of episode \
        trajectories containing experiences (obs,
                                actions,episode_reward)
    Returns:
        elite_batch_obs
        elite_batch_actions
        elite_reard_threshold


    """
    batch_obs, batch_actions,
    batch_rewards = zip(*trajectories)
    reward_threshold = np.percentile(batch_rewards,
                                elitism_criterion)
    indices = [index for index, value in enumerate(
            batch_rewards) if value >= reward_threshold]

    elite_batch_obs = [batch_obs[i] for i in indices]
    elite_batch_actions = [batch_actions[i] for i in \
                        indices]
    unpacked_elite_batch_obs = [item for items in \
                    elite_batch_obs for item in items]
    unpacked_elite_batch_actions = [item for items in \
                elite_batch_actions for item in items]
    return np.array(unpacked_elite_batch_obs), \
            np.array(unpacked_elite_batch_actions), \
            reward_threshold
```

9. Let's now test the elite experience gathering routine:

```
elite_obs, elite_actions, reward_threshold = gather_
elite_xp(trajectories, elitism_criterion=75)
```

10. Let's now look at implementing a helper method to convert discrete action indices to one-hot encoded vectors or probability distribution over actions:

```
def gen_action_distribution(action_index, action_dim=5):
    action_distribution = np.zeros(action_dim).\
                            astype(type(action_index))
    action_distribution[action_index] = 1
    action_distribution = \
                np.expand_dims(action_distribution, 0)
    return action_distribution
```

11. It's now time to test the action distribution generation function:

```
elite_action_distributions = np.array([gen_action_
distribution(a.item()) for a in elite_actions])
```

12. Now, let's create and compile the neural network brain with TensorFlow 2.x using the Keras functional API:

```
brain = Brain(env.action_space.n)
brain.compile(loss="categorical_crossentropy",
optimizer="adam", metrics=["accuracy"])
```

13. You can now test the brain training loop as follows:

```
elite_obs, elite_action_distributions = elite_obs.
astype("float16"), elite_action_distributions.
astype("float16")
brain.fit(elite_obs, elite_action_distributions, batch_
size=128, epochs=1);
```

This should produce the following output:

```
1/1 [==============================] - 0s 960us/step -
loss: 0.8060 - accuracy: 0.4900
```

> **Note**
>
> The numbers may vary.

14. The next big step is to implement an agent class that can be initialized with a brain to act in an environment:

```
class Agent(object):
    def __init__(self, brain):
        """Agent with a neural-network brain powered
           policy

        Args:
            brain (keras.Model): Neural Network based \
            model
        """
        self.brain = brain
        self.policy = self.policy_mlp

    def policy_mlp(self, observations):
        observations = observations.reshape(1, -1)
        action_logits = self.brain.process(observations)
        action = tf.random.categorical(
                tf.math.log(action_logits), num_samples=1)
        return tf.squeeze(action, axis=1)

    def get_action(self, observations):
        return self.policy(observations)
```

15. Next, we will implement a helper function to evaluate the agent in a given environment:

```
def evaluate(agent, env, render=True):
    obs, episode_reward, done, step_num = env.reset(),
                                        0.0, False, 0
    while not done:
        action = agent.get_action(obs)
        obs, reward, done, info = env.step(action)
        episode_reward += reward
```

```
        step_num += 1
        if render:
            env.render()
    return step_num, episode_reward, done, info
```

16. Let's now test the agent evaluation loop:

```
env = GridworldEnv()
agent = Agent(brain)
for episode in tqdm(range(10)):
    steps, episode_reward, done, info = evaluate(agent,
                                                 env)
env.close()
```

17. As a next step, let's define the parameters for the training loop:

```
total_trajectory_rollouts = 70
elitism_criterion = 70  # percentile
num_epochs = 200
mean_rewards = []
elite_reward_thresholds = []
```

18. Let's now create the environment, brain, and agent objects:

```
env = GridworldEnv()
input_shape = (env.observation_space.shape[0] * \
               env.observation_space.shape[1], )
brain = Brain(env.action_space.n)

brain.compile(loss="categorical_crossentropy",
              optimizer="adam", metrics=["accuracy"])
agent = Agent(brain)


for i in tqdm(range(num_epochs)):
    trajectories = [Trajectory(*rollout(agent, env)) \
            for _ in range(total_trajectory_rollouts)]
    _, _, batch_rewards = zip(*trajectories)
    elite_obs, elite_actions, elite_threshold = \
```

```
                    gather_elite_xp(trajectories,
                        elitism_criterion=elitism_criterion)
        elite_action_distributions = \
            np.array([gen_action_distribution(a.item()) \
                        for a in elite_actions])
        elite_obs, elite_action_distributions = \
            elite_obs.astype("float16"),
            elite_action_distributions.astype("float16")
        brain.fit(elite_obs, elite_action_distributions,
                batch_size=128, epochs=3, verbose=0);
        mean_rewards.append(np.mean(batch_rewards))
        elite_reward_thresholds.append(elite_threshold)
        print(f"Episode#:{i + 1} elite-reward-\
            threshold:{elite_reward_thresholds[-1]:.2f} \
            reward:{mean_rewards[-1]:.2f} ")

    plt.plot(mean_rewards, 'r', label="mean_reward")
    plt.plot(elite_reward_thresholds, 'g',
            label="elites_reward_threshold")
    plt.legend()
    plt.grid()
    plt.show()
```

This will generate a plot like the one shown in the following diagram:

> **Important note**
> The episode rewards will vary and the plots may look different.

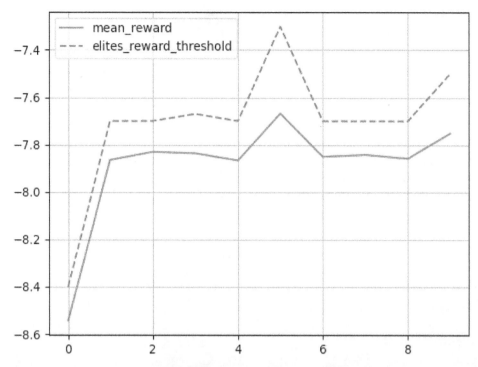

Figure 1.14 – Plot of the mean reward (solid, red) and reward threshold for elites (dotted, green)

The solid line in the plot is the mean reward obtained by the neural evolutionary agent, and the dotted line shows the reward threshold used for determining the elites.

How it works...

On every iteration, the evolutionary process rolls out or collects a bunch of trajectories to build up the experience data using the current set of neural weights in the agent's brain. An elite selection process is then employed that picks the top k percentile (elitism criterion) trajectories/experiences based on the episode reward obtained in that trajectory. This shortlisted experience data is then used to update the agent's brain model. The process repeats for a preset number of iterations allowing the agent's brain model to improve and collect more rewards.

See also

For more information, I suggest reading *The CMA Evolution Strategy: A Tutorial*: https://arxiv.org/pdf/1604.00772.pdf.

2
Implementing Value-Based, Policy-Based, and Actor-Critic Deep RL Algorithms

This chapter provides a practical approach to building value-based, policy-based, and actor-critic algorithm-based **reinforcement learning (RL)** agents. It includes recipes for implementing value iteration-based learning agents and breaks down the implementation details of several foundational algorithms in RL into simple steps. The policy gradient-based agent and the actor-critic agent make use of the latest major version of **TensorFlow 2.x** to define the neural network policies.

The following recipes will be covered in this chapter:

- Building stochastic environments for training RL agents
- Building value-based (RL) agent algorithms
- Implementing temporal difference learning
- Building Monte Carlo prediction and control algorithms for RL
- Implementing the SARSA algorithm and an RL agent
- Building a Q-learning agent
- Implementing policy gradients
- Implementing actor-critic algorithms

Let's get started!

Technical requirements

The code in this book has been tested extensively on Ubuntu 18.04 and Ubuntu 20.04, and should work with later versions of Ubuntu if Python 3.6+ is available. With Python 3.6 installed, along with the necessary Python packages listed at the beginning of each recipe, the code should run fine on Windows and Mac OS X too. It is advised that you create and use a Python virtual environment named `tf2rl-cookbook` to install the packages and run the code in this book. Installing Miniconda or Anaconda for Python virtual environment management is recommended.

The complete code for each recipe in each chapter is available here: `https://github.com/PacktPublishing/Tensorflow-2-Reinforcement-Learning-Cookbook`.

Building stochastic environments for training RL agents

To train RL agents for the real world, we need learning environments that are stochastic, since real-world problems are stochastic in nature. This recipe will walk you through the steps for building a **Maze** learning environment to train RL agents. The Maze is a simple, stochastic environment where the world is represented as a grid. Each location on the grid can be referred to as a cell. The goal of an agent in this environment is to find its way to the goal state. Consider the maze shown in the following diagram, where the black cells represent walls:

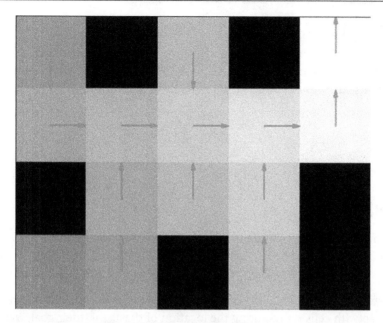

Figure 2.1 – The Maze environment

The agent's location is initialized to be at the top-left cell in the Maze. The agent needs to find its way around the grid to reach the goal located at the top-right cell in the Maze, collecting a maximum number of coins along the way while avoiding walls. The location of the goal, coins, walls, and the agent's starting location can be modified in the environment's code.

The four-dimensional discrete actions that are supported in this environment are as follows:

- *0*: Move up
- *1*: Move down
- *2*: Move left
- *3*: Move right

The reward is based on the number of coins that are collected by the agent before they reach the goal state. Because the environment is stochastic, the action that's taken by the environment has a slight (0.1) probability of "slipping" wherein the actual action that's executed will be altered stochastically. The slip action will be the clockwise directional action (LEFT -> UP, UP -> RIGHT, and so on). For example, with `slip_probability=0.2`, there is a 0.2 probability that a RIGHT action may result in DOWN.

Getting ready

To complete this recipe, you will need to activate the tf2rl-cookbook Python/conda virtual environment and run pip install -r requirements.txt. If the following import statements run without issues, you are ready to get started:

```
import gym
import numpy as np
```

Now, we can begin.

How to do it...

The learning environment is a simulator that provides observations for the RL agent, supports a set of actions that the RL agent can perform by executing the actions, and returns the resultant/new observation as a result of the agent taking the action.

Follow these steps to implement a stochastic Maze learning environment that represents a simple 2D map with cells representing the location of the agent, their goal, walls, coins, and empty space:

1. We'll start by defining the MazeEnv class and a map of the Maze environment:

    ```
    class MazeEnv(gym.Env):
        def __init__(self, stochastic=True):
            """Stochastic Maze environment with coins,\
                obstacles/walls and a goal state.
            """
            self.map = np.asarray(["SWFWG", "OOOOO", "WOOOW",
                                   "FOWFW"])
    ```

2. Next, place the obstacles/walls on the environment map in the appropriate places:

    ```
            self.dim = (4, 5)
            self.img_map = np.ones(self.dim)
            self.obstacles = [(0, 1), (0, 3), (2, 0),
                              (2, 4), (3, 2), (3, 4)]
            for x in self.obstacles:
                self.img_map[x[0]][x[1]] = 0
    ```

3. Let's define the slip mapping action in clockwise order:

```
self.slip_action_map = {
            0: 3,
            1: 2,
            2: 0,
            3: 1,
     }
```

4. Now, let's define a lookup table in the form of a dictionary to map indices to cells in the Maze environment:

```
self.index_to_coordinate_map = {
            0: (0, 0),
            1: (1, 0),
            2: (3, 0),
            3: (1, 1),
            4: (2, 1),
            5: (3, 1),
            6: (0, 2),
            7: (1, 2),
            8: (2, 2),
            9: (1, 3),
           10: (2, 3),
           11: (3, 3),
           12: (0, 4),
           13: (1, 4),
     }
```

5. Next, let's define the reverse lookup to find a cell, when given an index:

```
self.coordinate_to_index_map = dict((val, key) for \
     key, val in self.index_to_coordinate_map.items())
```

With that, we have finished initializing the environment!

6. Now, let's define a method that will handle the coins and their statuses in the Maze, where 0 means that the coin wasn't collected by the agent and 1 means that the coin was collected by the agent:

```python
def num2coin(self, n: int):
    coinlist = [
        (0, 0, 0),
        (1, 0, 0),
        (0, 1, 0),
        (0, 0, 1),
        (1, 1, 0),
        (1, 0, 1),
        (0, 1, 1),
        (1, 1, 1),
    ]
    return list(coinlist[n])
```

7. Now, let's define a quick method that will do the inverse operation of finding the number status/value of a coin:

```python
def coin2num(self, v: List):
    if sum(v) < 2:
        return np.inner(v, [1, 2, 3])
    else:
        return np.inner(v, [1, 2, 3]) + 1
```

8. Next, we will define a setter function to set the state of the environment. This is useful for algorithms such as value iteration, where each and every state needs to be visited in the environment for it to calculate values:

```python
def set_state(self, state: int) -> None:
    """Set the current state of the environment.
        Useful for value iteration

        Args:
            state (int): A valid state in the Maze env \
            int: [0, 112]
    """
    self.state = state
```

9. Now, it is time to implement the `step` method. We'll begin by implementing the `step` method and applying the `slip` action based on `slip_probability`:

```python
def step(self, action, slip=True):
    """Run one step into the Maze env

    Args:
        state (Any): Current index state of the maze
        action (int): Discrete action for up, down, \
        left, right
        slip (bool, optional): Stochasticity in the \
        env. Defaults to True.

    Raises:
        ValueError: If invalid action is provided as
        input

    Returns:
        Tuple : Next state, reward, done, _
    """
    self.slip = slip
    if self.slip:
        if np.random.rand() < self.slip_probability:
            action = self.slip_action_map[action]
```

10. Continuing with our implementation of the `step` function, we'll update the state of the maze based on the action that's taken:

```python
cell = self.index_to_coordinate_map[int(self.state / 8)]
if action == 0:
    c_next = cell[1]
    r_next = max(0, cell[0] - 1)
elif action == 1:
    c_next = cell[1]
    r_next = min(self.dim[0] - 1, cell[0] + 1)
elif action == 2:
    c_next = max(0, cell[1] - 1)
    r_next = cell[0]
```

```
        elif action == 3:
            c_next = min(self.dim[1] - 1, cell[1] + 1)
            r_next = cell[0]
        else:
            raise ValueError(f"Invalid action:{action}")
```

11. Next, we will determine whether the agent has reached the goal:

```
if (r_next == self.goal_pos[0]) and (
        c_next == self.goal_pos[1]
    ):  # Check if goal reached
        v_coin = self.num2coin(self.state % 8)
        self.state = 8 * self.coordinate_to_index_\
            map[(r_next, c_next)] + self.state % 8
    return (
        self.state,
        float(sum(v_coin)),
        True,
    )
```

12. Next, we'll handle cases when the action results in hitting an obstacle/wall:

```
else:
    if (r_next, c_next) in self.obstacles:  # obstacle
    # tuple list
            return self.state, 0.0, False
```

13. The last case you need to handle is seeing whether the action leads to collecting a coin:

```
else:  # Coin locations
            v_coin = self.num2coin(self.state % 8)
            if (r_next, c_next) == (0, 2):
                v_coin[0] = 1
            elif (r_next, c_next) == (3, 0):
                v_coin[1] = 1
            elif (r_next, c_next) == (3, 3):
                v_coin[2] = 1
            self.state = 8 * self.coordinate_to_
```

```
        index_map[(r_next, c_next)] + self.coin2num(v_coin)
                return (
                    self.state,
                    0.0,
                    False,
                )
```

14. To visualize the state of the Gridworld in a human-friendly manner, let's implement a render function that will print out a text version of the current state of the Maze environment:

```
def render(self):
        cell = self.index_to_coordinate_map[int(
                                    self.state / 8)]
        desc = self.map.tolist()

        desc[cell[0]] = (
            desc[cell[0]][: cell[1]]
            + "\x1b[1;34m"
            + desc[cell[0]][cell[1]]
            + "\x1b[0m"
            + desc[cell[0]][cell[1] + 1 :]
        )

        print("\n".join("".join(row) for row in desc))
```

15. To test whether the environment is working as expected, let's add a __main__ function that gets executed if the environment script is run directly:

```
if __name__ == "__main__":
    env = MazeEnv()
    obs = env.reset()
    env.render()
    done = False
    step_num = 1
    action_list = ["UP", "DOWN", "LEFT", "RIGHT"]
    # Run one episode
    while not done:
```

```
        # Sample a random action from the action space
        action = env.action_space.sample()
        next_obs, reward, done = env.step(action)
        print(
            f"step#:{step_num} action:\
            {action_list[action]} reward:{reward} \
            done:{done}"
        )
        step_num += 1
        env.render()
    env.close()
```

16. With that, we're all set! The Maze environment is ready and we can quickly test it by running the script (`python envs/maze.py`). An output similar to the following will be displayed:

Figure 2.2 – Textual representation of the Maze environment highlighting and underlining the agent's current state

Let's see how it works.

How it works...

Our `map`, as defined in *step 1* in the *How to do it...* section, represents the state of the learning environment. The Maze environment defines the observation space, the action space, and the rewarding mechanism for implementing a **Markov decision process (MDP)**. We sampled a valid action from the action space of the environment and stepped the environment with the chosen action, which resulted in us getting the new observation, reward, and a done status Boolean (representing whether the episode has finished) as the response from the Maze environment. The `env.render()` method converts the environment's internal grid representation into a simple text/string grid and prints it for easy visual understanding.

Building value-based reinforcement learning agent algorithms

Value-based reinforcement learning works by learning the state-value function or the action-value function in a given environment. This recipe will show you how to create and update the value function for the Maze environment to obtain an optimal policy. Learning value functions, especially in model-free RL problems where a model of the environment is not available, can prove to be quite effective, especially for RL problems with low-dimensional state space.

Upon completing this recipe, you will have an algorithm that can generate the following optimal action sequence based on value functions:

Figure 2.3 – Optimal action sequence generated by a value-based RL algorithm with state values represented through a jet color map

Let's get started.

Getting ready

To complete this recipe, you will need to activate the `tf2rl-cookbook` Python/conda virtual environment and run `pip install numpy gym`. If the following import statement runs without issues, you are ready to get started:

```
import numpy as np
```

Now, we can begin.

How to do it...

Let's implement a value function learning algorithm based on value iteration. We will use the Maze environment to implement and analyze the value iteration algorithm.

Follow these steps to implement this recipe:

1. Import the Maze learning environment from `envs.maze`:

   ```
   from envs.maze import MazeEnv
   ```

2. Create an instance of `MazeEnv` and print the observation space and action space:

   ```
   env = MazeEnv()
   print(f"Observation space: {env.observation_space}")
   print(f"Action space: {env.action_space}")
   ```

3. Let's define the state dimension in order to initialize `state-values`, `state-action values`, and our policy:

   ```
   state_dim = env.distinct_states
   state_values = np.zeros(state_dim)
   q_values = np.zeros((state_dim, env.action_space.n))
   policy = np.zeros(state_dim)
   ```

4. Now, we are ready to implement a function that can calculate the state/action value when given a state in the environment and an action. We will begin by declaring the `calculate_values` function; we'll complete the implementation in the following steps:

   ```
   def calculate_values(state, action):
       """Evaluate Value function for given state and action
   ```

```
Args:
    state (int): Valid (discrete) state in discrete \
    `env.observation_space`
    action (int): Valid (discrete) action in \
    `env.action_space`

Returns:
    v_sum: value for given state, action
"""
```

5. As the next step, we will generate `slip_action`, which is a stochastic action based on the stochasticity of the learning environment:

```
slip_action = env.slip_action_map[action]
```

6. When calculating the values of a given state-action pair, it is important to be able to set the state of the environment before executing an action to observe the reward/result. The Maze environment provides a convenient `set_state` method for setting the current state of the environment. Let's make use of it and step through the environment with the desired (input) action:

```
env.set_state(state)
slip_next_state, slip_reward, _ = \
                    env.step(slip_action, slip=False)
```

7. We need a list of transitions in the environment to be able to calculate the rewards, as per the Bellman equations. Let's create a `transitions` list and append the newly obtained environment transition information:

```
transitions = []
transitions.append((slip_reward, slip_next_state,
                    env.slip))
```

8. Let's obtain another transition using the state and the action, this time without stochasticity. We can do this by not using `slip_action` and setting `slip=False` while stepping through the Maze environment:

```
env.set_state(state)
next_state, reward, _ = env.step(action, slip=False)
transitions.append((reward, next_state,
                    1 - env.slip))
```

9. There is only one more step needed to complete the `calculate_values` function, which is to calculate the values:

```
for reward, next_state, pi in transitions:
    v_sum += pi * (reward + discount * \
                   state_values[next_state])
return v_sum
```

10. Now, we can start implementing the state/action value learning. We will begin by defining the `max_iteration` hyperparameters:

```
# Define the maximum number of iterations per learning
# step
max_iteration = 1000
```

11. Let's implement the `state-value` function learning loop using value iteration:

```
for i in range(iters):
    v_s = np.zeros(state_dim)
    for state in range(state_dim):
        if env.index_to_coordinate_map[int(state / 8)]==\
        env.goal_pos:
            continue
        v_max = float("-inf")
        for action in range(env.action_space.n):
            v_sum = calculate_values(state, action)
            v_max = max(v_max, v_sum)
        v_s[state] = v_max
    state_values = np.copy(v_s)
```

12. Now that we have the `state-value` function learning loop implemented, let's move on and implement the `action-value` function:

```
for state in range(state_dim):
    for action in range(env.action_space.n):
        q_values[state, action] = calculate_values(state,
                                                    action)
```

13. With the `action-value` function computed, we are only one step away from obtaining the optimal policy. Let's go get it!

```
for state in range(state_dim):
    policy[state] = np.argmax(q_values[state, :])
```

14. We can print the Q values (the `state-action` values) and the policy using the following lines of code:

```
print(f"Q-values: {q_values}")
print("Action mapping:[0 - UP; 1 - DOWN; 2 - LEFT; \
        3 - RIGHT")
print(f"optimal_policy: {policy}")
```

15. As a final step, let's visualize the value function's learning and policy updates:

```
from value_function_utils import viusalize_maze_values
viusalize_maze_values(q_values, env)
```

The preceding code will generate the following diagrams, which show the progress of the value function while it's learning and the policy updates:

Figure 2.4 – Progression (from left to right and from top to bottom) of the
learned value function and the policy

How it works...

The Maze environment contains a start cell, a goal cell, and a few cells containing coins, walls, and open spaces. There are 112 distinct states in the Maze environment due to the varying nature of the cells with coins. For illustration purposes, when an agent collects one of the coins, the environment is in a completely different state compared to the state when the agent collects a different coin. This is because the location of the coin also matters.

`q_values` (state-action values) is a big matrix of size 112 x 4, so it will print a long list of values. We will not show these here. The other two print statements in *step 14* should produce an output similar to the following:

```
Action mapping:[0 - UP; 1 - DOWN; 2 - LEFT; 3 - RIGHT
Optimal actions:
[1. 1. 1. 1. 1. 1. 1. 1. 3. 3. 3. 3. 3. 3. 3. 3. 1. 1. 3. 1. 3. 1. 3. 3.
 3. 1. 3. 3. 3. 3. 3. 3. 0. 3. 0. 0. 3. 0. 0. 0. 2. 2. 0. 2. 0. 2. 0. 0.
 0. 1. 0. 0. 1. 1. 0. 1. 0. 1. 0. 0. 3. 3. 0. 3. 0. 3. 0. 0. 3. 0. 0. 0.
 1. 1. 1. 2. 3. 3. 3. 3. 1. 1. 1. 0. 1. 0. 0. 0. 1. 1. 1. 0. 1. 0. 0. 0.
 1. 0. 0. 0. 0. 0. 0. 0. 2. 2. 2. 2. 0. 0. 0. 0.]
```

Figure 2.5 – Textual representation of the optimal action sequence

Value iteration-based value function learning follows Bellman equations, and the optimal policy is obtained from the Q-value function by simply choosing the action with the highest Q/action-value.

In *Figure 2.4*, the value function is represented using a jet color map, while the policy is represented using the green-arrows. Initially, the values for the states are almost even. As the learning progress, states with coins get more value than states without coins, and the state that leads to the goal gets a very high value that's only slightly less than the goal state itself. The black cells in the maze represents the walls. The arrows represent the directional action that the policy is prescribing from the given cell in the maze. As the learning converges, as shown in the bottom-right diagram, the policy is optimal, leading the agent to the goal after it's collected every coin.

> **Important note**
>
> The color versions of the diagrams in this book are available to download. You can find the link to these diagrams in the *Preface* of this book.

Implementing temporal difference learning

This recipe will walk you through how to implement the **temporal difference (TD)** learning algorithm. TD algorithms allow us to incrementally learn from incomplete episodes of agent experiences, which means they can be used for problems that require online learning capabilities. TD algorithms are useful in model-free RL settings as they do not depend on a model of the MDP transitions or rewards. To visually understand the learning progression of the TD algorithm, this recipe will also show you how to implement the GridworldV2 learning environment, which looks as follows when rendered:

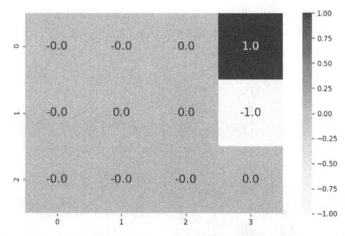

Figure 2.6 – The GridworldV2 learning environment 2D rendering with
state values and grid cell coordinates

Getting ready

To complete this recipe, you will need to activate the tf2rl-cookbook Python/conda virtual environment and run pip install numpy gym. If the following import statements run without issues, you are ready to get started:

```
import gym
import matplotlib.pyplot as plt
import numpy as np
```

Now, we can begin.

How to do it...

This recipe will contain two components that we will put together at the end. The first component is the GridworldV2 implementation, while the second component is the TD learning algorithm's implementation. Let's get started:

1. We will start by implementing GridworldV2 and then by defining the GridworldV2Eng class:

```python
class GridworldV2Env(gym.Env):
    def __init__(self, step_cost=-0.2, max_ep_length=500,
    explore_start=False):
        self.index_to_coordinate_map = {
            "0": [0, 0],
            "1": [0, 1],
            "2": [0, 2],
            "3": [0, 3],
            "4": [1, 0],
            "5": [1, 1],
            "6": [1, 2],
            "7": [1, 3],
            "8": [2, 0],
            "9": [2, 1],
            "10": [2, 2],
            "11": [2, 3],
        }
        self.coordinate_to_index_map = {
            str(val): int(key) for key, val in self.
    index_to_coordinate_map.items()
        }
```

2. In this step, you will continue implementing the __init__ method and define the necessary values that define the size of the Gridworld, the goal location, the wall location, and the location of the bomb, among other things:

```python
self.map = np.zeros((3, 4))
        self.observation_space = gym.spaces.Discrete(1)
        self.distinct_states = [str(i) for i in \
                                range(12)]
        self.goal_coordinate = [0, 3]
```

```
        self.bomb_coordinate = [1, 3]
        self.wall_coordinate = [1, 1]
        self.goal_state = self.coordinate_to_index_map[
                       str(self.goal_coordinate)]   # 3
        self.bomb_state = self.coordinate_to_index_map[
                       str(self.bomb_coordinate)]   # 7
        self.map[self.goal_coordinate[0]]\
               [self.goal_coordinate[1]] = 1
        self.map[self.bomb_coordinate[0]]\
               [self.bomb_coordinate[1]] = -1
        self.map[self.wall_coordinate[0]]\
               [self.wall_coordinate[1]] = 2

        self.exploring_starts = explore_start
        self.state = 8
        self.done = False
        self.max_ep_length = max_ep_length
        self.steps = 0
        self.step_cost = step_cost
        self.action_space = gym.spaces.Discrete(4)
        self.action_map = {"UP": 0, "RIGHT": 1,
                          "DOWN": 2, "LEFT": 3}
        self.possible_actions = \
                       list(self.action_map.values())
```

3. Now, we can move on to the definition of the reset() method, which will be
    called at the start of every episode, including the first one:

```
def reset(self):
        self.done = False
        self.steps = 0
        self.map = np.zeros((3, 4))
        self.map[self.goal_coordinate[0]]\
               [self.goal_coordinate[1]] = 1
        self.map[self.bomb_coordinate[0]]\
               [self.bomb_coordinate[1]] = -1
        self.map[self.wall_coordinate[0]]\
```

```
                      [self.wall_coordinate[1]] = 2


        if self.exploring_starts:
            self.state = np.random.choice([0, 1, 2, 4, 6,
                                           8, 9, 10, 11])
        else:
            self.state = 8
        return self.state
```

4. Let's implement a `get_next_state` method so that we can conveniently obtain the next state:

```
def get_next_state(self, current_position, action):

    next_state = self.index_to_coordinate_map[
                        str(current_position)].copy()

    if action == 0 and next_state[0] != 0 and \
    next_state != [2, 1]:
        # Move up
        next_state[0] -= 1
    elif action == 1 and next_state[1] != 3 and \
    next_state != [1, 0]:
        # Move right
        next_state[1] += 1
    elif action == 2 and next_state[0] != 2 and \
    next_state != [0, 1]:
        # Move down
        next_state[0] += 1
    elif action == 3 and next_state[1] != 0 and \
    next_state != [1, 2]:
        # Move left
        next_state[1] -= 1
    else:
        pass
    return self.coordinate_to_index_map[str(
                                next_state)]
```

5. With that, we are ready to implement the main `step` method of the `GridworldV2` environment:

```python
def step(self, action):
    assert action in self.possible_actions, \
        f"Invalid action:{action}"

    current_position = self.state
    next_state = self.get_next_state(
                        current_position, action)
    self.steps += 1
    if next_state == self.goal_state:
        reward = 1
        self.done = True
    elif next_state == self.bomb_state:
        reward = -1
        self.done = True
    else:
        reward = self.step_cost
    if self.steps == self.max_ep_length:
        self.done = True
    self.state = next_state
    return next_state, reward, self.done
```

6. Now, we can move on and implement the temporal difference learning algorithm. Let's begin by initializing the state values of the grid using a 2D numpy array and then set the value of the goal location and the bomb state:

```python
def temporal_difference_learning(env, max_episodes):
    grid_state_values = np.zeros((len(
                        env.distinct_states), 1))
    grid_state_values[env.goal_state] = 1
    grid_state_values[env.bomb_state] = -1
```

7. Next, let's define the discount factor, `gamma`, the learning rate, `alpha`, and initialize done to `False`:

```python
    # v: state-value function
    v = grid_state_values
```

```
gamma = 0.99  # Discount factor
alpha = 0.01  # learning rate
done = False
```

8. We can now define the main outer loop so that it runs `max_episodes` times, resetting the state of the environment to its initial state at the start of every episode:

```
for episode in range(max_episodes):
    state = env.reset()
```

9. Now, it's time to implement the inner loop with the temporal difference learning update one-liner:

```
while not done:
    action = env.action_space.sample()
        # random policy
    next_state, reward, done = env.step(action)

    # State-value function updates using TD(0)
    v[state] += alpha * (reward + gamma * \
                    v[next_state] - v[state])
    state = next_state
```

10. Once the learning has converged, we want to be able to visualize the state values for each state in the GridwordV2 environment. To do that, we can make use of the `visualize_grid_state_values` function from `value_function_utils`:

```
visualize_grid_state_values(grid_state_values.reshape((3,
4)))
```

11. We are now ready to run the `temporal_difference_learning` function from our main function:

```
if __name__ == "__main__":
    max_episodes = 4000
    env = GridworldV2Env(step_cost=-0.1,
                    max_ep_length=30)
    temporal_difference_learning(env, max_episodes)
```

12. The preceding code will take a few seconds to run temporal difference learning for `max_episodes`. It will then produce a diagram similar to the following:

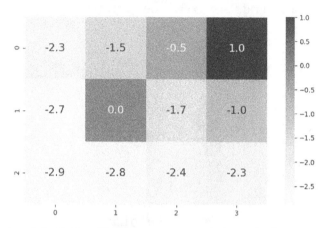

Figure 2.7 – Rendering of the GridworldV2 environment, with the grid cell coordinates and state values colored according to the scale shown on the right

How it works...

Based on our environment's implementation, you may have noticed that `goal_state` is located at (0, 3) and that `bomb_state` is located at (1, 3). This is based on the coordinates, colors, and values of the grid cells:

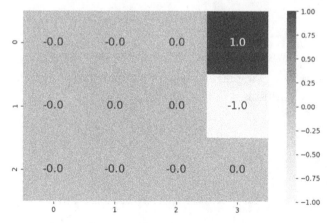

Figure 2.8 – Rendering of the GridWorldV2 environment with initial state values

The state is linearized and is represented using a single integer indicating each of the 12 distinct states in the GridWorldV2 environment. The following diagram shows a linearized rendering of the grid states to give you a better understanding of the state encoding:

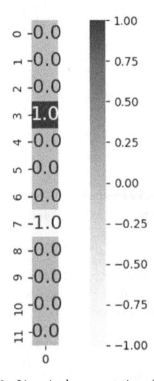

Figure 2.9 – Linearized representation of the states

Now that we have seen how to implement temporal difference learning, let's move on to building Monte Carlo algorithms.

Building Monte Carlo prediction and control algorithms for RL

This recipe provides the ingredients for building a **Monte Carlo** prediction and control algorithm so that you can build your RL agents. Similar to the temporal difference learning algorithm, Monte Carlo learning methods can be used to learn both the state and the action value functions. Monte Carlo methods have zero bias since they learn from complete episodes with real experience, without approximate predictions. These methods are suitable for applications that require good convergence properties. The following diagram illustrates the value that's learned by the Monte Carlo method for the GridworldV2 environment:

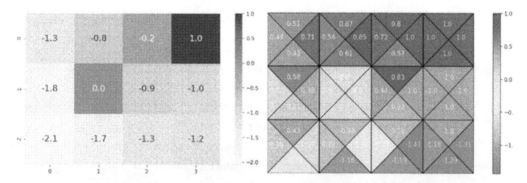

Figure 2.10 – Monte Carlo prediction of state values (left) and state-action values (right)

Getting ready

To complete this recipe, you will need to activate the tf2rl-cookbook Python/conda virtual environment and run pip install -r requirements.txt. If the following import statement runs without issues, you are ready to get started:

```
import numpy as np
```

Now, let's begin.

How to do it...

We will start by implementing the monte_carlo_prediction algorithm and visualizing the learned value function for each state in the GridworldV2 environment. After that, we will implement an **epsilon-greedy policy** and the monte_carlo_control algorithm to construct an agent that will act in an RL environment.

Follow these steps:

1. Let's start with the import statements and import the necessary Python modules:

    ```python
    import numpy as np


    from envs.gridworldv2 import GridworldV2Env
    from value_function_utils import (
        visualize_grid_action_values,
        visualize_grid_state_values,
    )
    ```

2. The next step is to define the `monte_carlo_prediction` function and initialize the necessary objects, as shown here:

    ```python
    def monte_carlo_prediction(env, max_episodes):
        returns = {state: [] for state in \
                    env.distinct_states}
        grid_state_values = np.zeros(len(
                                env.distinct_states))
        grid_state_values[env.goal_state] = 1
        grid_state_values[env.bomb_state] = -1
        gamma = 0.99  # Discount factor
    ```

3. Now, let's implement the outer loop. Outer loops are commonplace in all RL agent training code:

    ```python
    for episode in range(max_episodes):
        g_t = 0
        state = env.reset()
        done = False
        trajectory = []
    ```

4. Next up is the inner loop:

```
while not done:
    action = env.action_space.sample()
    # random policy
    next_state, reward, done = env.step(action)
    trajectory.append((state, reward))
    state = next_state
```

5. We now have all the information we need to compute the state values of the states in the grid:

```
for idx, (state, reward) in enumerate(trajectory[::-1]):
    g_t = gamma * g_t + reward
    # first visit Monte-Carlo prediction
    if state not in np.array(trajectory[::-1])\
    [:, 0][idx + 1 :]:
        returns[str(state)].append(g_t)
        grid_state_values[state] =
np.mean(returns[str(state)])
```

Let's visualize the learned state value function using the `visualize_grid_state_values` helper function from the `value_function_utils` script:

```
visualize_grid_state_values(grid_state_values.reshape((3, 4)))
```

6. Now, it's time to run our Monte Carlo predictor:

```
if __name__ == "__main__":
    max_episodes = 4000
    env = GridworldV2Env(step_cost=-0.1,
                         max_ep_length=30)
    print(f"===Monte Carlo Prediction===")
    monte_carlo_prediction(env, max_episodes)
```

7. The preceding code should produce a diagram showing the rendering for the GridworldV2 environment, along with state values:

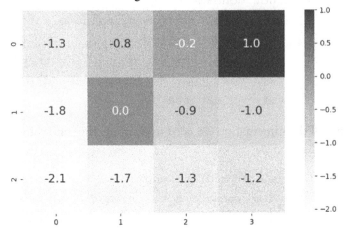

Figure 2.11 – Rendering of GridworldV2 with state values learned using the Monte Carlo prediction algorithm

8. Let's implement a function for the epsilon-greedy policy:

```python
def epsilon_greedy_policy(action_logits, epsilon=0.2):
    idx = np.argmax(action_logits)
    probs = []
    epsilon_decay_factor = np.sqrt(sum([a ** 2 for a in \
                            action_logits]))

    if epsilon_decay_factor == 0:
        epsilon_decay_factor = 1.0
    for i, a in enumerate(action_logits):
        if i == idx:
            probs.append(round(1 - epsilon + (
                    epsilon / epsilon_decay_factor), 3))
        else:
            probs.append(round(
                    epsilon / epsilon_decay_factor, 3))
    residual_err = sum(probs) - 1
    residual = residual_err / len(action_logits)

    return np.array(probs) - residual
```

9. Now, let's move on to the implementation of the **Monte Carlo Control** algorithm for reinforcement learning. We will start by defining the function, along with the initial values for the state-action values:

```
def monte_carlo_control(env, max_episodes):
    grid_state_action_values = np.zeros((12, 4))
    grid_state_action_values[3] = 1
    grid_state_action_values[7] = -1
```

10. Let's continue with the implementation of the Monte Carlo Control function by initializing the returns for all the possible state and action pairs:

```
    possible_states = ["0", "1", "2", "3", "4", "5", "6",
"7", "8", "9", "10", "11"]
    possible_actions = ["0", "1", "2", "3"]
    returns = {}
    for state in possible_states:
        for action in possible_actions:
            returns[state + ", " + action] = []
```

11. As the next step, let's define the outer loop for each episode and then the inner loop for each step in an episode. By doing this, we can collect trajectories of experience until the end of an episode:

```
gamma = 0.99
    for episode in range(max_episodes):
        g_t = 0
        state = env.reset()
        trajectory = []
        while True:
            action_values = \
                grid_state_action_values[state]
            probs = epsilon_greedy_policy(action_values)
            action = np.random.choice(np.arange(4), \
                                p=probs)  # random policy

            next_state, reward, done = env.step(action)
            trajectory.append((state, action, reward))
```

```
        state = next_state
        if done:
            break
```

12. Now that we have a full trajectory for an episode in the inner loop, we can implement our Monte Carlo Control update to update the state-action values:

```
for step in reversed(trajectory):
    g_t = gamma * g_t + step[2]
    Returns[str(step[0]) + ", " + \
            str(step[1])].append(g_t)
    grid_state_action_values[step[0]][step[1]]= \
    np.mean(
        Returns[str(step[0]) + ", " + \
            str(step[1])]
    )
```

13. Once the outer loop completes, we can visualize the state-action values using the `visualize_grid_action_values` helper function from the `value_function_utils` script:

```
visualize_grid_action_values(grid_state_action_values
```

14. Finally, let's run our `monte_carlo_control` function to learn the state-action values in the GridworldV2 environment and display the learned values:

```
if __name__ == "__main__":
    max_episodes = 4000
    env = GridworldV2Env(step_cost=-0.1, \
                         max_ep_length=30)
    print(f"===Monte Carlo Control===")
    monte_carlo_control(env, max_episodes)
```

The preceding code will produce a rendering similar to the following:

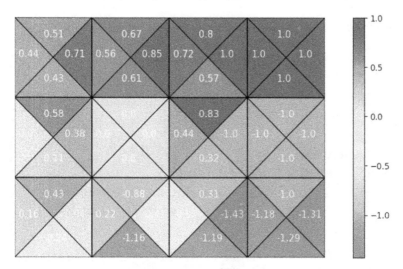

Figure 2.12 – Rendering of the GridworldV2 environment with four action values per grid state shown using rectangles

That concludes this recipe!

How it works...

Monte Carlo methods for episodic tasks learn directly from experience from full sample returns obtained in an episode. The Monte Carlo prediction algorithm for estimating the value function based on first visit averaging is as follows:

Initialize:
    $\pi \leftarrow$ policy to be evaluated
    $V \leftarrow$ an arbitrary state-value function
    $Returns(s) \leftarrow$ an empty list, for all $s \in \mathcal{S}$

Repeat forever:
    Generate an episode using π
    For each state s appearing in the episode:
        $G \leftarrow$ return following the first occurrence of s
        Append G to $Returns(s)$
        $V(s) \leftarrow$ average($Returns(s)$)

Figure 2.13 – Monte Carlo prediction algorithm

Once a series of trajectories have been collected by the agent, we can use the transition information in the Monte Carlo Control algorithm to learn the state-action value function. This can be used by an agent so that they can act in a given RL environment.

The Monte Carlo Control algorithm is shown in the following diagram:

Initialize, for all $s \in \mathcal{S}$, $a \in \mathcal{A}(s)$:
$Q(s,a) \leftarrow$ arbitrary
$Returns(s,a) \leftarrow$ empty list
$\pi \leftarrow$ an arbitrary ε-soft policy

Repeat forever:
(a) Generate an episode using π
(b) For each pair s, a appearing in the episode:
$R \leftarrow$ return following the first occurrence of s, a
Append R to $Returns(s,a)$
$Q(s,a) \leftarrow$ average$(Returns(s,a))$
(c) For each s in the episode:
$a^* \leftarrow \arg\max_a Q(s,a)$
For all $a \in \mathcal{A}(s)$:
$$\pi(s,a) \leftarrow \begin{cases} 1 - \varepsilon + \varepsilon/|\mathcal{A}(s)| & \text{if } a = a^* \\ \varepsilon/|\mathcal{A}(s)| & \text{if } a \neq a^* \end{cases}$$

Figure 2.14 – Monte-Carlo Control algorithm

The results of the learned state-action value function are shown in *Figure 2.12*, where each triangle in a grid cell shows the state-action value of taking that directional action in that grid state. The base of the triangle lies in the direction of the action. For example, the triangle in the top-left corner of *Figure 2.12* that has a value of 0.44 is the state-action value of taking the LEFT action in that grid state.

Implementing the SARSA algorithm and an RL agent

This recipe will show you how to implement the **State-Action-Reward-State-Action (SARSA)** algorithm, as well as how to develop and train an agent using the SARSA algorithm so that it can act in a reinforcement learning environment. The SARSA algorithm can be applied to model-free control problems and allows us to optimize the value function of an unknown MDP.

Upon completing this recipe, you will have a working RL agent that, when acting in the GridworldV2 environment, will generate the following state-action value function using the SARSA algorithm:

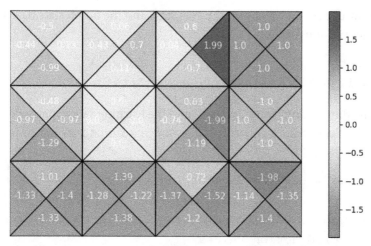

Figure 2.15 – Rendering of the GridworldV2 environment – each triangle represents the action value of taking that directional action in that grid state

Getting ready

To complete this recipe, you will need to activate the `tf2rl-cookbook` Python/conda virtual environment and run `pip install -r requirements.txt`. If the following import statements run without issues, you are ready to get started:

```
import numpy as np
import random
```

Now, let's begin.

How to do it...

Let's implement the SARSA learning update as a function and make use of an epsilon-greedy exploration policy. With these two pieces combined, we will have a complete agent to act in a given RL environment. In this recipe, we will train and test the agent in the GridworldV2 environment.

Let's start our implementation step by step:

1. First, let's define a function for implementing the SARSA algorithm and initialize the state-action values with zeros:

```
def sarsa(env, max_episodes):
    grid_action_values = np.zeros((len(
                  env.distinct_states), env.action_space.n))
```

2. We can now update the values for the goal state and the bomb state based on the environment's configuration:

```
grid_action_values[env.goal_state] = 1
grid_action_values[env.bomb_state] = -1
```

3. Let's define the discount factor, gamma, and the learning rate hyperparameter, alpha. Also, let's create a convenient alias for grid_action_values by calling it q:

```
gamma = 0.99  # discounting factor
alpha = 0.01  # learning rate
# q: state-action-value function
q = grid_action_values
```

4. Let's begin to implement the outer loop:

```
for episode in range(max_episodes):
    step_num = 1
    done = False
    state = env.reset()
    action = greedy_policy(q[state], 1)
```

5. Now, it's time to implement the inner loop with the SARSA learning update step:

```
while not done:
        next_state, reward, done = env.step(action)
        step_num += 1
        decayed_epsilon = gamma ** step_num
        # Doesn't have to be gamma
        next_action = greedy_policy(q[next_state], \
                                    decayed_epsilon)
        q[state][action] += alpha * (
            reward + gamma * q[next_state] \
                [next_action] - q[state][action]
        )
        state = next_state
        action = next_action
```

6. As the final step in the `sarsa` function, let's visualize the state-action value function:

```
visualize_grid_action_values(grid_action_values)
```

7. Now, we will implement the epsilon-greedy policy that the agent will use:

```
def greedy_policy(q_values, epsilon):
    """Epsilon-greedy policy """

    if random.random() >= epsilon:
        return np.argmax(q_values)
    else:
        return random.randint(0, 3)
```

8. Finally, we must implement the main function and run the SARSA algorithm:

```
if __name__ == "__main__":
    max_episodes = 4000
    env = GridworldV2Env(step_cost=-0.1, \
                         max_ep_length=30)
    sarsa(env, max_episodes)
```

When executed, a rendering of the GridworldV2 environment with the state-action values will appear, as shown in the following diagram:

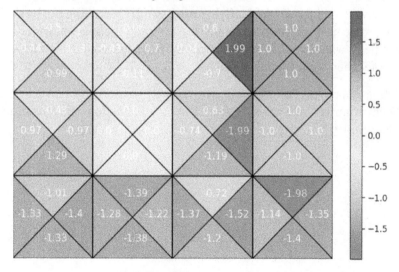

Figure 2.16 – Output of the SARSA algorithm in the GridworldV2 environment

How it works...

SARSA is an on-policy temporal difference learning-based control algorithm. This recipe made uses of the SARSA algorithm to estimate the optimal state-action values. The SARSA algorithm can be summarized as follows:

Initialize $Q(s, a), \forall s \in \mathcal{S}, a \in \mathcal{A}(s)$, arbitrarily, and $Q(terminal\text{-}state, \cdot) = 0$
Repeat (for each episode):
    Initialize S
    Choose A from S using policy derived from Q (e.g., ϵ-greedy)
    Repeat (for each step of episode):
        Take action A, observe R, S'
        Choose A' from S' using policy derived from Q (e.g., ϵ-greedy)
        $Q(S, A) \leftarrow Q(S, A) + \alpha \left[R + \gamma Q(S', A') - Q(S, A) \right]$
        $S \leftarrow S'; A \leftarrow A';$
    until S is terminal

Figure 2.17 – SARSA algorithm

As you may be able to tell, this is very similar to the Q-learning algorithm. The similarities will become clear when we look at the next recipe in this chapter, *Building a Q-learning agent*.

Building a Q-learning agent

This recipe will show you how to build a **Q-learning** agent. Q-learning can be applied to model-free RL problems. It supports off-policy learning and therefore provides a practical solution to problems where available experiences were/are collected using some other policy or by some other agent (even humans).

Upon completing this recipe, you will have a working RL agent that, when acting in the GridworldV2 environment, will generate the following state-action value function using the SARSA algorithm:

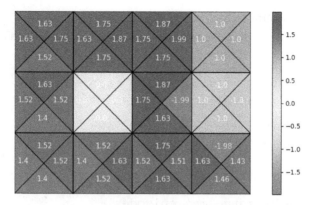

Figure 2.18 – State-action values obtained using the Q-learning algorithm

Getting ready

To complete this recipe, you will need to activate the tf2rl-cookbook Python/conda virtual environment and run `pip install -r requirements.txt`. If the following import statements run without issues, you are ready to get started:

```
import numpy as np
import random
```

Now, let's begin.

How to do it...

Let's implement the Q-learning algorithm as a function, as well as an epsilon-greedy policy to build our Q-learning agent.

Let's start our implementation:

1. First, let's define a function for implementing the Q-learning algorithm and initialize the state-action values with zeros:

    ```
    def q_learning(env, max_episodes):
        grid_action_values = np.zeros((len(\
            env.distinct_states), env.action_space.n))
    ```

2. We can now update the values for the goal state and the bomb state based on the environment's configuration:

    ```
    grid_action_values[env.goal_state] = 1
    grid_action_values[env.bomb_state] = -1
    ```

3. Let's define the discount factor, gamma, and the learning rate hyperparameter, alpha. Also, let's create a convenient alias for `grid_action_values` by calling it q:

```
gamma = 0.99  # discounting factor
alpha = 0.01  # learning rate
# q: state-action-value function
q = grid_action_values
```

4. Let's begin to implement the outer loop:

```
for episode in range(max_episodes):
    step_num = 1
    done = False
    state = env.reset()
```

5. As the next step, let's implement the inner loop with the Q-learning update. We will also decay the epsilon used in the epsilon-greedy policy:

```
while not done:
    decayed_epsilon = 1 * gamma ** step_num
    # Doesn't have to be gamma
    action = greedy_policy(q[state], \
            decayed_epsilon)
    next_state, reward, done = env.step(action)

    # Q-Learning update
    grid_action_values[state][action] += alpha * (
        reward + gamma * max(q[next_state]) - \
        q[state][action]
    )
    step_num += 1
    state = next_state
```

6. As the final step in the `q_learning` function, let's visualize the state-action value function:

```
visualize_grid_action_values(grid_action_values)
```

7. Next, we will implement the epsilon-greedy policy that the agent will use:

```python
def greedy_policy(q_values, epsilon):
    """Epsilon-greedy policy """

    if random.random() >= epsilon:
        return np.argmax(q_values)
    else:
        return random.randint(0, 3)
```

8. Finally, we will implement the main function and run the SARSA algorithm:

```python
if __name__ == "__main__":
    max_episodes = 4000
    env = GridworldV2Env(step_cost=-0.1,
                         max_ep_length=30)
    q_learning(env, max_episodes)
```

When executed, a rendering of the GridworldV2 environment with the state-action values will appear, as shown in the following diagram:

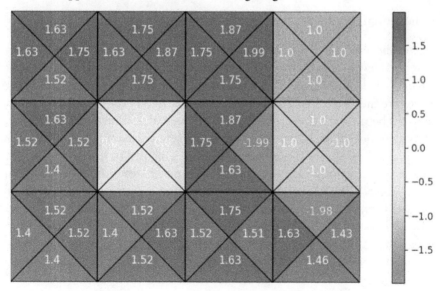

Figure 2.19 – Rendering of the GridworldV2 environment with the action values obtained using the Q-learning algorithm

How it works...

The Q-learning algorithm involves the Q value update, which can be summarized by the following equation:

$$Q[s, a] = Q[s, a] + \lambda * (r + \gamma * max_{a'}(Q[s', a']) - Q[s, a])$$

Here, we have the following:

- $Q[s, a]$ is the value of the Q function for the current state, s, and action, a.
- $max(Q\_[s\_, a\_]) - Q[s, a])$ is used for choosing the maximum value from the possible next steps.
- s is the current position of the agent.
- a is the current action.
- λ is the learning rate.
- r is the reward that is received in the current position.
- γ is the gamma (reward decay, discount factor).
- s' is the next state.
- a' is the actions that are available in the next state, s'.

As you may now be able to tell, the difference between Q-learning and SARSA is only in how the action-value/Q-value of the next state and action pair is calculated. In Q-learning, we use $max(Q[s', a'])$, the maximum value of the Q-function, whereas in the SARSA algorithm, we take the Q-value of the action that was chosen in the next state. This may sound subtle, but because the Q-learning algorithm infers the value by taking the max over all actions and doesn't just infer based on the current behavior policy, it can directly learn the optimal policy. On the other hand, the SARSA algorithm learns a near-optimal policy based on the behavior policy's exploration parameter (for example, the ε parameter in the ε-greedy policy). The SARSA algorithm has a better convergence property than the Q-learning algorithm, so it is more suited for cases where learning happens online and or on a real-world system, or even if there are real resources (time and/or money) being spent compared to training in a simulation or simulated worlds. Q-learning is more suited for training an "optimal" agent in simulation or when the resources (like time/money) are not too costly.

Implementing policy gradients

Policy gradient algorithms are fundamental to reinforcement learning and serve as the basis for several advanced RL algorithms. These algorithms directly optimize for the best policy, which can lead to faster learning compared to value-based algorithms. Policy gradient algorithms are effective for problems/applications with high-dimensional or continuous action spaces. This recipe will show you how to implement policy gradient algorithms using TensorFlow 2.0. Upon completing this recipe, you will be able to train an RL agent in any compatible OpenAI Gym environment.

Getting ready

To complete this recipe, you will need to activate the `tf2rl-cookbook` Python/conda virtual environment and run `pip install -r requirements.txt`. If the following import statements run without issues, you are ready to get started:

```
import tensorflow as tf
import tensorflow_probability as tfp
from tensorflow import keras
from tensorflow.keras import layers
import numpy as np
import gym
```

Now, let's begin.

How to do it...

There are three main parts to this recipe. The first one is applying the policy function, which is going to be represented using a neural network implemented in TensorFlow 2.x. The second part is applying the Agent class' implementation, while the final part will be to apply a trainer function, which is used to train the policy gradient-based agent in a given RL environment.

Let's start implementing the parts one by one:

1. The first step is to define the `PolicyNet` class. We will define the model so that it has three fully connected or **dense** neural network layers:

```
class PolicyNet(keras.Model):
    def __init__(self, action_dim=1):
        super(PolicyNet, self).__init__()
        self.fc1 = layers.Dense(24, activation="relu")
```

```
    self.fc2 = layers.Dense(36, activation="relu")
    self.fc3 = layers.Dense(action_dim,
                              activation="softmax")
```

2. Next, we will implement the `call` function, which will be called to process inputs to the model:

```
def call(self, x):
    x = self.fc1(x)
    x = self.fc2(x)
    x = self.fc3(x)
    return x
```

3. Let's also define a `process` function that we can call with a batch of observations to be processed by the model:

```
def process(self, observations):
    # Process batch observations using `call(x)`
    # behind-the-scenes
    action_probabilities = \
        self.predict_on_batch(observations)
    return action_probabilities
```

4. With the policy network defined, we can implement the `Agent` class, which utilizes the policy network, and an optimizer for training the model:

```
class Agent(object):
    def __init__(self, action_dim=1):
        """Agent with a neural-network brain powered
            policy

        Args:
            action_dim (int): Action dimension
        """
        self.policy_net = PolicyNet(
                            action_dim=action_dim)
        self.optimizer = tf.keras.optimizers.Adam(
                            learning_rate=1e-3)
        self.gamma = 0.99
```

5. Now, let's define a policy helper function that takes an observation as input, has it processed by the policy network, and returns the action as the output:

```
def policy(self, observation):
    observation = observation.reshape(1, -1)
    observation = tf.convert_to_tensor(observation,
                                    dtype=tf.float32)
    action_logits = self.policy_net(observation)
    action = tf.random.categorical(
            tf.math.log(action_logits), num_samples=1)
    return action
```

6. Let's define another helper function to get the action from the agent:

```
def get_action(self, observation):
    action = self.policy(observation).numpy()
    return action.squeeze()
```

7. Now, it's time to define the learning updates for the policy gradient algorithm. Let's initialize the learn function with an empty list for discounted rewards:

```
def learn(self, states, rewards, actions):
    discounted_reward = 0
    discounted_rewards = []
    rewards.reverse()
```

8. This is the right place to calculate the discounted rewards while using the episodic rewards as input:

```
for r in rewards:
    discounted_reward = r + self.gamma * \
                            discounted_reward
    discounted_rewards.append(discounted_reward)
    discounted_rewards.reverse()
```

9. Now, let's implement the crucial step of calculating the policy gradient and update the parameters of the neural network policy using an optimizer:

```
for state, reward, action in zip(states,
discounted_rewards, actions):
    with tf.GradientTape() as tape:
```

```
                action_probabilities = \
                    self.policy_net(np.array([state]),\
                                    training=True)
                loss = self.loss(action_probabilities, \
                                 action, reward)
            grads = tape.gradient(loss,
                    self.policy_net.trainable_variables)
            self.optimizer.apply_gradients(
                zip(grads,
                    self.policy_net.trainable_variables)
            )
```

10. Let's implement the loss function that we referred to in the previous step to calculate the policy parameter updates:

```
def loss(self, action_probabilities, action, reward):
    dist = tfp.distributions.Categorical(
        probs=action_probabilities, dtype=tf.float32
    )
    log_prob = dist.log_prob(action)
    loss = -log_prob * reward
    return loss
```

11. With the Agent class fully implemented, we can move on to implementing the agent training function. Let's start with the function's definition:

```
def train(agent: Agent, env: gym.Env, episodes: int,
render=True):
    """Train `agent` in `env` for `episodes`

    Args:
        agent (Agent): Agent to train
        env (gym.Env): Environment to train the agent
        episodes (int): Number of episodes to train
        render (bool): True=Enable/False=Disable \
                        rendering; Default=True
    """
```

12. Now, let's begin with the outer loop implementation of the agent training function:

```
for episode in range(episodes):
        done = False
        state = env.reset()
        total_reward = 0
        rewards = []
        states = []
        actions = []
```

13. Let's continue to implement the inner loop to finalize the `train` function:

```
        while not done:
            action = agent.get_action(state)
            next_state, reward, done, _ = \
                                env.step(action)
            rewards.append(reward)
            states.append(state)
            actions.append(action)
            state = next_state
            total_reward += reward
            if render:
                env.render()
            if done:
                agent.learn(states, rewards, actions)
                print("\n")
            print(f"Episode#:{episode} \
            ep_reward:{total_reward}", end="\r")
```

14. Finally, we need to implement the main function:

```
if __name__ == "__main__":
    agent = Agent()
    episodes = 5000
    env = gym.make("MountainCar-v0")
    train(agent, env, episodes)
    env.close()
```

The preceding code will launch the training process for the agent in the **MountainCar** environment. This will render the environment (since `render=True`) and display what the agent is doing in the environment with respect to driving the car uphill. Once the agent has been trained for a sufficient number of episodes, you will see the agent driving the car all the way up hill, as shown in the following diagram:

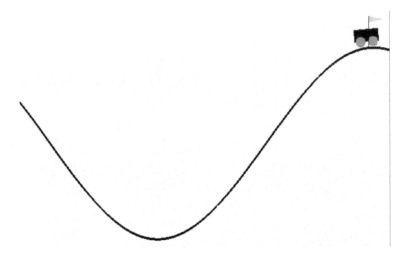

Figure 2.20 – Policy gradient agent completing the MountainCar task

That concludes this recipe!

How it works...

We used TensorFlow 2.x's **Keras API** to define a multilayer feed-forward neural network model that represents the RL agent's policy. We then defined an Agent class that utilizes the neural network policy to act in the `MountainCar` RL environment. The policy gradient algorithm is shown in the following diagram:

Input: a differentiable policy parameterization $\pi(a|s, \boldsymbol{\theta}), \forall a \in \mathcal{A}, s \in \mathcal{S}, \boldsymbol{\theta} \in \mathbb{R}^n$
Initialize policy weights $\boldsymbol{\theta}$
Repeat forever:
    Generate an episode $S_0, A_0, R_1, \dots, S_{T-1}, A_{T-1}, R_T$, following $\pi(\cdot|\cdot, \boldsymbol{\theta})$
    For each step of the episode $t = 0, \dots, T - 1$:
        $G_t \leftarrow$ return from step t
        $\boldsymbol{\theta} \leftarrow \boldsymbol{\theta} + \alpha \gamma^t G_t \nabla_{\boldsymbol{\theta}} \log \pi(A_t|S_t, \boldsymbol{\theta})$

Figure 2.21 – Policy gradient algorithm

As you train the policy gradient-based agent, you will observe that while the agent can learn to drive the car up the mountain, this can take a long time or they may get stuck in local minima. This basic version of the policy gradient has some limitations. The policy gradient is an on-policy algorithm that can only use experiences/trajectories or episode transitions from the same policy that is being optimized. The basic version of the policy gradient algorithm does not provide a guarantee for monotonic improvements in performance as it can get stuck in local minima.

Implementing actor-critic RL algorithms

Actor-critic algorithms allow us to combine value-based and policy-based reinforcement learning – an all-in-one agent. While policy gradient methods directly search and optimize the policy in the policy space, leading to smoother learning curves and improvement guarantees, they tend to get stuck at the local maxima (for a long-term reward optimization objective). Value-based methods do not get stuck at local optimum values, but they lack convergence guarantees, and algorithms such as Q-learning tend to have high variance and are not very sample-efficient. Actor-critic methods combine the good qualities of both value-based and policy gradient-based algorithms. Actor-critic methods are also more sample-efficient. This recipe will make it easy for you to implement an actor-critic-based RL agent using TensorFlow 2.x. Upon completing this recipe, you will be able to train the actor-critic agent in any OpenAI Gym-compatible reinforcement learning environment. As an example, we will train the agent in the CartPole-V0 environment.

Getting ready

To complete this recipe, you will need to activate the tf2rl-cookbook Python/conda virtual environment and run pip install -r requirements.txt. If the following import statements run without issues, you are ready to get started:

```
import numpy as np
import tensorflow as tf
import gym
import tensorflow_probability as tfp
```

Now, let's begin.

How to do it...

There are three main parts to this recipe. The first is creating the actor-critic model, which is going to be represented using a neural network implemented in TensorFlow 2.x. The second part is creating the Agent class' implementation, while the final part is going to be about creating a trainer function that will train the policy gradient-based agent in a given RL environment.

Let's start implementing the parts one by one:

1. Let's begin with our implementation of the `ActorCritic` class:

```
class ActorCritic(tf.keras.Model):
    def __init__(self, action_dim):
        super().__init__()
        self.fc1 = tf.keras.layers.Dense(512, \
                                    activation="relu")
        self.fc2 = tf.keras.layers.Dense(128, \
                                    activation="relu")
        self.critic = tf.keras.layers.Dense(1, \
                                    activation=None)
        self.actor = tf.keras.layers.Dense(action_dim, \
                                    activation=None)
```

2. The final thing we need to do in the `ActorCritic` class is implement the `call` function, which performs a forward pass through the neural network model:

```
def call(self, input_data):
    x = self.fc1(input_data)
    x1 = self.fc2(x)
    actor = self.actor(x1)
    critic = self.critic(x1)
    return critic, actor
```

3. With the `ActorCritic` class defined, we can move on and implement the `Agent` class and initialize an `ActorCritic` model, along with an optimizer to update the parameters of the actor-critic model:

```
class Agent:
    def __init__(self, action_dim=4, gamma=0.99):
        """Agent with a neural-network brain powered
```

```
        policy

    Args:
        action_dim (int): Action dimension
        gamma (float) : Discount factor. Default=0.99
    """

    self.gamma = gamma
    self.opt = tf.keras.optimizers.Adam(
                            learning_rate=1e-4)
    self.actor_critic = ActorCritic(action_dim)
```

4. Next, we must implement the agent's `get_action` method:

```
def get_action(self, state):
    _, action_probabilities = \
                self.actor_critic(np.array([state]))
    action_probabilities = tf.nn.softmax(
                            action_probabilities)
    action_probabilities = \
                    action_probabilities.numpy()
    dist = tfp.distributions.Categorical(
        probs=action_probabilities, dtype=tf.float32
    )
    action = dist.sample()
    return int(action.numpy()[0])
```

5. Now, let's implement a function that will calculate the actor loss based on the actor-critic algorithm. This will drive the parameters of the actor-critic network and allow the agent to improve:

```
def actor_loss(self, prob, action, td):
    prob = tf.nn.softmax(prob)
    dist = tfp.distributions.Categorical(probs=prob,
                            dtype=tf.float32)
    log_prob = dist.log_prob(action)
    loss = -log_prob * td
    return loss
```

6. We are now ready to implement the learning function of the actor-critic agent:

```python
def learn(self, state, action, reward, next_state, done):
    state = np.array([state])
    next_state = np.array([next_state])

    with tf.GradientTape() as tape:
        value, action_probabilities = \
            self.actor_critic(state, training=True)
        value_next_st, _ = self.actor_critic(
                        next_state, training=True)
        td = reward + self.gamma * value_next_st * \
            (1 - int(done)) - value
        actor_loss = self.actor_loss(
                    action_probabilities, action, td)
        critic_loss = td ** 2
        total_loss = actor_loss + critic_loss
    grads = tape.gradient(total_loss,
            self.actor_critic.trainable_variables)
    self.opt.apply_gradients(zip(grads,
            self.actor_critic.trainable_variables))
    return total_loss
```

7. Now, let's define the training function for training the agent in a given RL environment:

```python
def train(agent, env, episodes, render=True):
    """Train `agent` in `env` for `episodes`

    Args:
        agent (Agent): Agent to train
        env (gym.Env): Environment to train the agent
        episodes (int): Number of episodes to train
        render (bool): True=Enable/False=Disable \
                        rendering; Default=True
    """
    for episode in range(episodes):
        done = False
```

```
    state = env.reset()
    total_reward = 0
    all_loss = []
    while not done:
        action = agent.get_action(state)
        next_state, reward, done, _ = \
                                    env.step(action)
        loss = agent.learn(state, action, reward,
                            next_state, done)
        all_loss.append(loss)
        state = next_state
        total_reward += reward
        if render:
            env.render()
        if done:
            print("\n")
        print(f"Episode#:{episode}
                ep_reward:{total_reward}",
            end="\r")
```

8. The final step is to implement the main function, which will call the trainer to train the agent for the specified number of episodes:

```
if __name__ == "__main__":

    env = gym.make("CartPole-v0")
    agent = Agent(env.action_space.n)
    num_episodes = 20000
    train(agent, env, num_episodes)
```

Once the agent has been sufficiently trained, you will see that the agent is able to balance the pole on the cart pretty well, as shown in the following diagram:

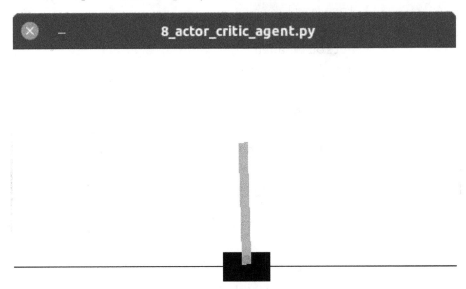

Figure 2.22 – Actor-critic agent solving the CartPole task

How it works...

In this recipe, we defined a neural network-based actor-critic model using TensorFlow 2.x's Keras API. In the neural network model, we defined two fully connected or dense neural network layers to extract features from the input. This produced two outputs corresponding to the output for an actor and an output for the critic. The critic's output is a single float value, whereas the actor's output represents the logits for each of the allowed actions in a given RL environment.

3
Implementing Advanced RL Algorithms

This chapter provides short and crisp recipes to implement advanced **Reinforcement Learning (RL)** algorithms and agents from scratch using **TensorFlow 2.x**. It includes recipes to build **Deep-Q-Networks (DQN)**, **Double and Dueling Deep Q-Networks (DDQN, DDDQN)**, **Deep Recurrent Q-Networks (DRQN)**, **Asynchronous Advantage Actor-Critic (A3C)**, **Proximal Policy Optimization (PPO)**, and **Deep Deterministic Policy Gradients (DDPG)**.

The following recipes are discussed in this chapter:

- Implementing the Deep Q-Learning algorithm, DQN, and Double-DQN agent
- Implementing the Dueling DQN agent
- Implementing the Dueling Double DQN algorithm and DDDQN agent
- Implementing the Deep Recurrent Q-Learning algorithm and DRQN agent
- Implementing the Asynchronous Advantage Actor-Critic algorithm and A3C agent
- Implementing the Proximal Policy Optimization algorithm and PPO agent
- Implementing the Deep Deterministic Policy Gradient algorithm and DDPG agent

Technical requirements

The code in the book is extensively tested on Ubuntu 18.04 and Ubuntu 20.04 and should work with later versions of Ubuntu if Python 3.6+ is available. With Python 3.6+ installed along with the necessary Python packages as listed before the start of each of the recipes, the code should run fine on Windows and Mac OS X too. It is advised to create and use a Python virtual environment named `tf2rl-cookbook` to install the packages and run the code in this book. Miniconda or Anaconda installation for Python virtual environment management is recommended.

The complete code for each recipe in each chapter is available here: `https://github.com/PacktPublishing/Tensorflow-2-Reinforcement-Learning-Cookbook`.

Implementing the Deep Q-Learning algorithm, DQN, and Double-DQN agent

DQN agent uses a deep neural network to learn the Q-value function. DQN has shown itself to be a powerful algorithm for discrete action-space environments and problems and is considered to be a notable milestone in the history of deep reinforcement learning when DQN mastered Atari Games.

The Double-DQN agent uses two identical deep neural networks that are updated differently and so hold different weights. The second neural network is a copy of the main neural network from some time in the past (typically from the last episode).

By the end of this recipe, you will have implemented a complete DQN and Double-DQN agent from scratch using TensorFlow 2.x that is ready to be trained in any discrete action-space RL environment.

Let's get started.

Getting ready

To complete this recipe, you will first need to activate the `tf2rl-cookbook` Conda Python virtual environment and `pip install -r requirements.txt`. If the following import statements run without issues, you are ready to get started!

```
import argparse
from datetime import datetime
import os
import random
from collections import deque

import gym
import numpy as np
import tensorflow as tf
from tensorflow.keras.layers import Dense, Input
```

Now we can begin.

How to do it...

The DQN agent comprises a few components, namely, the **Replay Buffer**, the DQN class, the Agent class, and the `train` method. Perform the following steps to implement each of these components from scratch to build a complete DQN agent using TensorFlow 2.x:

1. First, let's create an argument parser to handle configuration inputs to the script:

```
        parser = argparse.ArgumentParser(prog="TFRL-
Cookbook-Ch3-DQN")
parser.add_argument("--env , default="CartPole-v0")
parser.add_argument("--lr", type=float, default=0.005)
parser.add_argument("--batch_size", type=int,
default=256)
parser.add_argument("--gamma", type=float, default=0.95)
parser.add_argument("--eps", type=float, default=1.0)
```

```python
parser.add_argument("--eps_decay", type=float,
default=0.995)
parser.add_argument("--eps_min", type=float,
default=0.01)
parser.add_argument("--logdir", default="logs")
args = parser.parse_args()
```

2. Let's now create a Tensorboard logger to log useful statistics during the agent's training:

```python
logdir = os.path.join(
    args.logdir, parser.prog, args.env,
    datetime.now().strftime("%Y%m%d-%H%M%S")
)
print(f"Saving training logs to:{logdir}")
writer = tf.summary.create_file_writer(logdir)
```

3. Next, let's implement a `ReplayBuffer` class:

```python
class ReplayBuffer:
    def __init__(self, capacity=10000):
        self.buffer = deque(maxlen=capacity)

    def store(self, state, action, reward, next_state,
    done):
        self.buffer.append([state, action, reward,
        next_state, done])

    def sample(self):
        sample = random.sample(self.buffer,
                                args.batch_size)
        states, actions, rewards, next_states, done = \
                            map(np.asarray, zip(*sample))
        states = np.array(states).reshape(
                                args.batch_size, -1)
        next_states = np.array(next_states).\
                            reshape(args.batch_size, -1)
        return states, actions, rewards, next_states,
```

```
            done

    def size(self):
        return len(self.buffer)
```

4. It's now time to implement the DQN class that defines the deep neural network in TensorFlow 2.x:

```
class DQN:
    def __init__(self, state_dim, aciton_dim):
        self.state_dim = state_dim
        self.action_dim = aciton_dim
        self.epsilon = args.eps

        self.model = self.nn_model()

    def nn_model(self):
        model = tf.keras.Sequential(
            [
                Input((self.state_dim,)),
                Dense(32, activation="relu"),
                Dense(16, activation="relu"),
                Dense(self.action_dim),
            ]
        )
        model.compile(loss="mse",
                optimizer=Adam(args.lr))
        return model
```

5. To get the prediction and action from the DQN, let's implement the predict and get_action methods:

```
    def predict(self, state):
        return self.model.predict(state)

    def get_action(self, state):
        state = np.reshape(state, [1, self.state_dim])
        self.epsilon *= args.eps_decay
```

```
        self.epsilon = max(self.epsilon, args.eps_min)
        q_value = self.predict(state)[0]
        if np.random.random() < self.epsilon:
            return random.randint(0, self.action_dim - 1)
        return np.argmax(q_value)
    def train(self, states, targets):
        self.model.fit(states, targets, epochs=1)
```

6. With the other components implemented, we can begin implementing our
    Agent class:

```
class Agent:
    def __init__(self, env):
        self.env = env
        self.state_dim = \
            self.env.observation_space.shape[0]
        self.action_dim = self.env.action_space.n

        self.model = DQN(self.state_dim, self.action_dim)
        self.target_model = DQN(self.state_dim,
                                self.action_dim)
        self.update_target()

        self.buffer = ReplayBuffer()

    def update_target(self):
        weights = self.model.model.get_weights()
        self.target_model.model.set_weights(weights)
```

7. The crux of the Deep Q-learning algorithm is the q-learning update and experience
    replay. Let's implement that next:

```
    def replay_experience(self):
        for _ in range(10):
            states, actions, rewards, next_states, done=\
                self.buffer.sample()
            targets = self.target_model.predict(states)
            next_q_values = self.target_model.\
```

```
                    predict(next_states).max(axis=1)
            targets[range(args.batch_size), actions] = (
                rewards + (1 - done) * next_q_values * \
                args.gamma
            )
            self.model.train(states, targets)
```

8. The next crucial step is to implement the `train` function to train the agent:

```
def train(self, max_episodes=1000):
    with writer.as_default():  # Tensorboard logging
        for ep in range(max_episodes):
            done, episode_reward = False, 0
            observation = self.env.reset()
            while not done:
                action = \
                    self.model.get_action(observation)
                next_observation, reward, done, _ = \
                    self.env.step(action)
                self.buffer.store(
                    observation, action, reward * \
                    0.01, next_observation, done
                )
                episode_reward += reward
                observation = next_observation
                if self.buffer.size() >= args.batch_size:
                    self.replay_experience()
                self.update_target()
                print(f"Episode#{ep} Reward:{
                                episode_reward}")
                tf.summary.scalar("episode_reward",
                                episode_reward, step=ep)
            writer.flush()
```

9. Finally, let's create the main function to start training the agent:

```
if __name__ == "__main__":
    env = gym.make("CartPole-v0")
```

```
agent = Agent(env)
agent.train(max_episodes=20000)
```

10. To train the DQN agent in the default environment (CartPole-v0), execute the following command:

```
python ch3-deep-rl-agents/1_dqn.py
```

11. You can also train the DQN agent in any OpenAI Gym-compatible discrete action-space environment using the command-line arguments:

```
python ch3-deep-rl-agents/1_dqn.py -env "MountainCar-v0"
```

12. Now, to implement the Double DQN agent, we must modify the replay_ experience method to use Double Q-learning's update step, as shown here:

```
def replay_experience(self):
    for _ in range(10):
        states, actions, rewards, next_states, done=\
            self.buffer.sample()
        targets = self.target_model.predict(states)
        next_q_values = \
            self.target_model.predict(next_states)[
            range(args.batch_size),
            np.argmax(self.model.predict(
                                next_states), axis=1),
        ]
        targets[range(args.batch_size), actions] = (
            rewards + (1 - done) * next_q_values * \
            args.gamma
        )
        self.model.train(states, targets)
```

13. Finally, to train the Double DQN agent, save and run the script with the updated replay_experience method or use the script provided as part of the source code for this book:

```
python ch3-deep-rl-agents/1_double_dqn.py
```

Let's see how it works.

How it works...

Updates to the weights in the DQN are performed as per the following Q learning equation:

$$\Delta w = \alpha \, [\, \underbrace{(R + \gamma max_a \hat{Q}(s', a; w))}_{Max\ Qvalue\ for\ s'} - \underbrace{\hat{Q}(s, a; w)}_{Predicted\ Qvalue}\,] \, \underbrace{\nabla_w \hat{Q}(s, a;\ w)}_{Gradient\ of\ Qvalue}$$

Here, Δw is the change in the parameters (weights) of the DQN, s is the current state, a is the current action, s' is the next state, w represents the weights of the DQN, γ is the discount factor, α is the learning rate, and $\hat{Q}(s, a;\ w)$ represents the Q-value for the given state (s) and action (a) predicted by the DQN with a weight w.

To understand the difference between the DQN agent and the Double-DQN agent, compare the `replay_experience` method in step 8 (DQN) and step 13 (Double DQN). You will notice that the key difference lies in calculating the `next_q_values`. The DQN agent uses the maximum of the predicted Q-values (which can be an overestimation), whereas the Double DQN agent uses the predicted Q-value using two distinct neural Q networks. This is done in Double DQN to avoid the problem of overestimating the Q-values by the DQN agent.

Implementing the Dueling DQN agent

A Dueling DQN agent explicitly estimates two quantities through a modified network architecture:

- State values, $V(s)$
- Advantage values, $A(s, a)$

The state value estimates the value of being in state s, and the advantage value represents the advantage of taking action a in state s. This key idea of explicitly and separately estimating the two quantities enables the Dueling DQN to perform better in comparison to DQN. This recipe will walk you through the steps to implement a Dueling DQN agent from scratch using TensorFlow 2.x.

Getting ready

To complete this recipe, you will first need to activate the `tf2rl-cookbook` Conda Python virtual environment and `pip install -r requirements.txt`. If the following import statements run without issues, you are ready to get started!

```
import argparse
import os
```

```
import random
from collections import deque
from datetime import datetime


import gym
import numpy as np
import tensorflow as tf
from tensorflow.keras.layers import Add, Dense, Input
from tensorflow.keras.optimizers import Adam
```

Now we can begin.

How to do it...

The Dueling DQN agent comprises a few components, namely, the **Replay Buffer**, the DuelingDQN class, the Agent class, and the train method. Perform the following steps to implement each of these components from scratch to build a complete Dueling DQN agent using TensorFlow 2.x:

1. As a first step, let's create an argument parser to handle command-line configuration inputs to the script:

```
parser = argparse.ArgumentParser(prog="TFRL-Cookbook-Ch3-
DuelingDQN")
parser.add_argument("--env", default="CartPole-v0")
parser.add_argument("--lr", type=float, default=0.005)
parser.add_argument("--batch_size", type=int, default=64)
parser.add_argument("--gamma", type=float, default=0.95)
parser.add_argument("--eps", type=float, default=1.0)
parser.add_argument("--eps_decay", type=float,
default=0.995)
parser.add_argument("--eps_min", type=float,
default=0.01)
parser.add_argument("--logdir", default="logs")

args = parser.parse_args()
```

2. To log useful statistics during the agent's training, let's create a TensorBoard logger:

```python
logdir = os.path.join(
    args.logdir, parser.prog, args.env,
    datetime.now().strftime("%Y%m%d-%H%M%S")
)
print(f"Saving training logs to:{logdir}")
writer = tf.summary.create_file_writer(logdir)
```

3. Next, let's implement a `ReplayBuffer` class:

```python
class ReplayBuffer:
    def __init__(self, capacity=10000):
        self.buffer = deque(maxlen=capacity)

    def store(self, state, action, reward, next_state,
    done):
        self.buffer.append([state, action, reward,
                            next_state, done])

    def sample(self):
        sample = random.sample(self.buffer,
                               args.batch_size)
        states, actions, rewards, next_states, done = \
                        map(np.asarray, zip(*sample))
        states = np.array(states).reshape(
                                args.batch_size, -1)
        next_states = np.array(next_states).reshape(
                                args.batch_size, -1)
        return states, actions, rewards, next_states,
        done

    def size(self):
        return len(self.buffer)
```

4. It's time to implement the DuelingDQN class that defines the deep neural network in TensorFlow 2.x:

```python
class DuelingDQN:
    def __init__(self, state_dim, aciton_dim):
        self.state_dim = state_dim
        self.action_dim = aciton_dim
        self.epsilon = args.eps


        self.model = self.nn_model()


    def nn_model(self):
        backbone = tf.keras.Sequential(
            [
                Input((self.state_dim,)),
                Dense(32, activation="relu"),
                Dense(16, activation="relu"),
            ]
        )
        state_input = Input((self.state_dim,))
        backbone_1 = Dense(32, activation="relu")\
                            (state_input)
        backbone_2 = Dense(16, activation="relu")\
                            (backbone_1)
        value_output = Dense(1)(backbone_2)
        advantage_output = Dense(self.action_dim)\
                                (backbone_2)
        output = Add()([value_output, advantage_output])
        model = tf.keras.Model(state_input, output)
        model.compile(loss="mse",
                        optimizer=Adam(args.lr))
        return model
```

5. To get the prediction and action from the Dueling DQN, let's implement the predict, get_action, and train methods:

```python
    def predict(self, state):
        return self.model.predict(state)


def get_action(self, state):
    state = np.reshape(state, [1, self.state_dim])
    self.epsilon *= args.eps_decay
    self.epsilon = max(self.epsilon, args.eps_min)
    q_value = self.predict(state)[0]
    if np.random.random() < self.epsilon:
        return random.randint(0, self.action_dim - 1)
    return np.argmax(q_value)


def train(self, states, targets):
    self.model.fit(states, targets, epochs=1)
```

6. We can now begin implementing our Agent class:

```python
class Agent:
    def __init__(self, env):
        self.env = env
        self.state_dim = \
            self.env.observation_space.shape[0]
        self.action_dim = self.env.action_space.n

        self.model = DuelingDQN(self.state_dim,
                                self.action_dim)
        self.target_model = DuelingDQN(self.state_dim,
                                       self.action_dim)
        self.update_target()

        self.buffer = ReplayBuffer()

    def update_target(self):
        weights = self.model.model.get_weights()
        self.target_model.model.set_weights(weights)
```

7. The crux of the Dueling Deep Q-learning algorithm is the q-learning update and experience replay. Let's implement that next:

```
def replay_experience(self):
    for _ in range(10):
        states, actions, rewards, next_states, done=\
            self.buffer.sample()
        targets = self.target_model.predict(states)
        next_q_values = self.target_model.\
                    predict(next_states).max(axis=1)
        targets[range(args.batch_size), actions] = (
            rewards + (1 - done) * next_q_values * \
            args.gamma
        )
        self.model.train(states, targets)
```

8. The next crucial step is to implement the `train` function to train the agent:

```
def train(self, max_episodes=1000):
    with writer.as_default():
        for ep in range(max_episodes):
            done, episode_reward = False, 0
            state = self.env.reset()
            while not done:
                action = self.model.get_action(state)
                next_state, reward, done, _ = \
                                self.env.step(action)
                self.buffer.put(state, action, \
                                reward * 0.01, \
                                next_state, done)
                episode_reward += reward
                state = next_state

            if self.buffer.size() >= args.batch_size:
                self.replay_experience()
            self.update_target()
            print(f"Episode#{ep} \
                Reward:{episode_reward}")
```

```
                    tf.summary.scalar("episode_reward",\
                                episode_reward, step=ep)
```

9. Finally, let's create the main function to start training the agent:

```
if __name__ == "__main__":
    env = gym.make("CartPole-v0")
    agent = Agent(env)
    agent.train(max_episodes=20000)
```

10. To train the Dueling DQN agent in the default environment (CartPole-v0), execute the following command:

```
python ch3-deep-rl-agents/2_dueling_dqn.py
```

11. You can also train the DQN agent in any OpenAI Gym-compatible discrete action-space environment using the command-line arguments:

```
python ch3-deep-rl-agents/2_dueling_dqn.py -env
"MountainCar-v0"
```

Let's see how it works.

How it works...

The Dueling-DQN agent differs from the DQN agent in terms of the neural network architecture.

The differences are summarized in the following diagram:

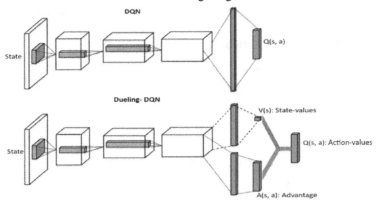

Figure 3.1 – DQN and Dueling-DQN compared

The DQN (top half of the diagram) has a linear architecture and predicts a single quantity (Q(s, a)), whereas the Dueling-DQN has a bifurcation in the last layer and predicts multiple quantities.

Implementing the Dueling Double DQN algorithm and DDDQN agent

Dueling Double DQN (DDDQN) combines the benefits of both Double Q-learning and Dueling architecture. Double Q-learning corrects DQN from overestimating the action values. The Dueling architecture uses a modified architecture to separately learn the state value function (V) and the advantage function (A). This explicit separation allows the algorithm to learn faster, especially when there are many actions to choose from and when the actions are very similar to each other. The dueling architecture enables the agent to learn even when only one action in a state has been taken, as it can update and estimate the state value function, unlike the DQN agent, which cannot learn from actions that were not taken yet. By the end of this recipe, you will have a complete implementation of the DDDQN agent.

Getting ready

To complete this recipe, you will first need to activate the `tf2rl-cookbook` Conda Python virtual environment and `pip install -r requirements.txt`. If the following import statements run without issues, you are ready to get started!

```
import argparse
from datetime import datetime
```

```
import os
import random
from collections import deque

import gym
import numpy as np
import tensorflow as tf
from tensorflow.keras.layers import Add, Dense, Input
from tensorflow.keras.optimizers import Adam
```

We are ready to begin!

How to do it...

The DDDQN agent combines the ideas in DQN, Double DQN and the Dueling DQN. Perform the following steps to implement each of these components from scratch to build a complete Dueling Double DQN agent using TensorFlow 2.x:

1. First, let's create an argument parser to handle configuration inputs to the script:

    ```
    parser = argparse.ArgumentParser(prog="TFRL-Cookbook-Ch3-
    DuelingDoubleDQN")
    parser.add_argument("--env", default="CartPole-v0")
    parser.add_argument("--lr", type=float, default=0.005)
    parser.add_argument("--batch_size", type=int,
    default=256)
    parser.add_argument("--gamma", type=float, default=0.95)
    parser.add_argument("--eps", type=float, default=1.0)
    parser.add_argument("--eps_decay", type=float,
    default=0.995)
    parser.add_argument("--eps_min", type=float,
    default=0.01)
    parser.add_argument("--logdir", default="logs")

    args = parser.parse_args()
    ```

2. Next, let's create a Tensorboard logger to log useful statistics during the agent's training process:

```
logdir = os.path.join(
    args.logdir, parser.prog, args.env, \
    datetime.now().strftime("%Y%m%d-%H%M%S")
)
print(f"Saving training logs to:{logdir}")
writer = tf.summary.create_file_writer(logdir)
```

3. Now, let's implement a `ReplayBuffer`:

```
class ReplayBuffer:
    def __init__(self, capacity=10000):
        self.buffer = deque(maxlen=capacity)

    def store(self, state, action, reward, next_state,
done):
        self.buffer.append([state, action, reward, \
        next_state, done])

    def sample(self):
        sample = random.sample(self.buffer, \
                                args.batch_size)
        states, actions, rewards, next_states, done = \
                                map(np.asarray, zip(*sample))
        states = np.array(states).reshape(
                                args.batch_size, -1)
        next_states = np.array(next_states).\
                            reshape(args.batch_size, -1)
        return states, actions, rewards, next_states, \
        done

    def size(self):
        return len(self.buffer)
```

4. It's now time to implement the Dueling DQN class that defines the neural network as per the dueling architecture to which we will add Double DQN updates in later steps:

```python
class DuelingDQN:
    def __init__(self, state_dim, aciton_dim):
        self.state_dim = state_dim
        self.action_dim = aciton_dim
        self.epsilon = args.eps


        self.model = self.nn_model()


    def nn_model(self):
        state_input = Input((self.state_dim,))
        fc1 = Dense(32, activation="relu")(state_input)
        fc2 = Dense(16, activation="relu")(fc1)
        value_output = Dense(1)(fc2)
        advantage_output = Dense(self.action_dim)(fc2)
        output = Add()([value_output, advantage_output])
        model = tf.keras.Model(state_input, output)
        model.compile(loss="mse", \
                        optimizer=Adam(args.lr))
        return model
```

5. To get the prediction and action from the Dueling DQN, let's implement the predict and get_action methods:

```python
    def predict(self, state):
        return self.model.predict(state)


    def get_action(self, state):
        state = np.reshape(state, [1, self.state_dim])
        self.epsilon *= args.eps_decay
        self.epsilon = max(self.epsilon, args.eps_min)
        q_value = self.predict(state)[0]
        if np.random.random() < self.epsilon:
            return random.randint(0, self.action_dim - 1)
```

```
            return np.argmax(q_value)


    def train(self, states, targets):
        self.model.fit(states, targets, epochs=1)
```

6. With the other components implemented, we can begin implementing our `Agent` class:

```
class Agent:
    def __init__(self, env):
        self.env = env
        self.state_dim = \
            self.env.observation_space.shape[0]
        self.action_dim = self.env.action_space.n

        self.model = DuelingDQN(self.state_dim,
                                self.action_dim)
        self.target_model = DuelingDQN(self.state_dim,
                                       self.action_dim)
        self.update_target()

        self.buffer = ReplayBuffer()

    def update_target(self):
        weights = self.model.model.get_weights()
        self.target_model.model.set_weights(weights)
```

7. The main elements in the Dueling Double Deep Q-learning algorithm are the Q-learning update and experience replay. Let's implement that next:

```
    def replay_experience(self):
        for _ in range(10):
            states, actions, rewards, next_states, done=\
                                    self.buffer.sample()
            targets = self.target_model.predict(states)
            next_q_values = \
                self.target_model.predict(next_states)[
```

```
                    range(args.batch_size),
                    np.argmax(self.model.predict(
                                    next_states), axis=1),
            ]
            targets[range(args.batch_size), actions] = (
                    rewards + (1 - done) * next_q_values * \
                    args.gamma
            )
            self.model.train(states, targets)
```

8. The next crucial step is to implement the `train` function to train the agent:

```
def train(self, max_episodes=1000):
        with writer.as_default():
            for ep in range(max_episodes):
                done, episode_reward = False, 0
                observation = self.env.reset()
                while not done:
                    action = \
                        self.model.get_action(observation)
                    next_observation, reward, done, _ = \
                        self.env.step(action)
                    self.buffer.store(
                        observation, action, reward * \
                            0.01, next_observation, done
                    )
                    episode_reward += reward
                    observation = next_observation

                    if self.buffer.size() >= args.batch_size:
                        self.replay_experience()
                self.update_target()
                print(f"Episode#{ep} \
                        Reward:{episode_reward}")
                tf.summary.scalar("episode_reward",
                                        episode_reward,
                                        step=ep)
```

9.   Finally, let's create the main function to start training the agent:

```
if __name__ == "__main__":
    env = gym.make("CartPole-v0")
    agent = Agent(env)
    agent.train(max_episodes=20000)
```

10.   To train the DQN agent in the default environment (CartPole-v0), execute the following command:

```
python ch3-deep-rl-agents/3_dueling_double_dqn.py
```

11.   You can also train the Dueling Double DQN agent in any OpenAI Gym-compatible discrete action-space environment using the command-line arguments:

```
python ch3-deep-rl-agents/3_dueling_double_dqn.py -env
"MountainCar-v0"
```

How it works...

The Dueling Double DQN architecture combines the advancements introduced by the Double DQN and Dueling architectures together.

Implementing the Deep Recurrent Q-Learning algorithm and DRQN agent

DRQN uses a recurrent neural network to learn the Q-value function. DRQN is more suited for reinforcement learning in environments with partial observability. The recurrent network layers in the DRQN allow the agent to learn by integrating information from a temporal sequence of observations. For example, DRQN agents can infer the velocity of moving objects in the environment without any changes to their inputs (for example, no frame stacking is required). By the end of this recipe, you will have a complete DRQN agent ready to be trained in an RL environment of your choice.

Getting ready

To complete this recipe, you will first need to activate the tf2rl-cookbook Conda Python virtual environment and pip install -r requirements.txt. If the following import statements run without issues, you are ready to get started!

```python
import tensorflow as tf
from datetime import datetime
import os
from tensorflow.keras.layers import Input, Dense, LSTM
from tensorflow.keras.optimizers import Adam
import gym
import argparse
import numpy as np
from collections import deque
import random
```

Let's begin!

How to do it...

The Dueling Double DQN agent combines the ideas in DQN, Double DQN, and the Dueling DQN. Perform the following steps to implement each of these components from scratch to build a complete DRQN agent using TensorFlow 2.x:

1. First, create an argument parser to handle configuration inputs to the script:

```python
parser = argparse.ArgumentParser(prog="TFRL-Cookbook-Ch3-DRQN")
parser.add_argument("--env", default="CartPole-v0")
parser.add_argument("--lr", type=float, default=0.005)
parser.add_argument("--batch_size", type=int, default=64)
parser.add_argument("--time_steps", type=int, default=4)
parser.add_argument("--gamma", type=float, default=0.95)
parser.add_argument("--eps", type=float, default=1.0)
parser.add_argument("--eps_decay", type=float,
default=0.995)
parser.add_argument("--eps_min", type=float,
default=0.01)
parser.add_argument("--logdir", default="logs")
args = parser.parse_args()
```

2. Let's log useful statistics during the agent's training using Tensorboard:

```python
logdir = os.path.join(
    args.logdir, parser.prog, args.env, \
```

```
            datetime.now().strftime("%Y%m%d-%H%M%S")
    )
    print(f"Saving training logs to:{logdir}")
    writer = tf.summary.create_file_writer(logdir)
```

3. Next, let's implement a `ReplayBuffer`:

```python
class ReplayBuffer:
    def __init__(self, capacity=10000):
        self.buffer = deque(maxlen=capacity)

    def store(self, state, action, reward, next_state,\
    done):
        self.buffer.append([state, action, reward, \
                            next_state, done])

    def sample(self):
        sample = random.sample(self.buffer,
                                args.batch_size)
        states, actions, rewards, next_states, done = \
            map(np.asarray, zip(*sample))
        states = np.array(states).reshape(
                                args.batch_size, -1)
        next_states = np.array(next_states).reshape(
                                args.batch_size, -1)
        return states, actions, rewards, next_states, \
        done

    def size(self):
        return len(self.buffer)
```

4. It's now time to implement the DRQN class that defines the deep neural network using TensorFlow 2.x:

```python
class DRQN:
    def __init__(self, state_dim, action_dim):
        self.state_dim = state_dim
        self.action_dim = action_dim
```

```
        self.epsilon = args.eps

        self.opt = Adam(args.lr)
        self.compute_loss = \
            tf.keras.losses.MeanSquaredError()
        self.model = self.nn_model()

    def nn_model(self):
        return tf.keras.Sequential(
            [
                Input((args.time_steps, self.state_dim)),
                LSTM(32, activation="tanh"),
                Dense(16, activation="relu"),
                Dense(self.action_dim),
            ]
        )
```

5. To get the prediction and action from the DRQN, let's implement the `predict` and `get_action` methods:

```
    def predict(self, state):
        return self.model.predict(state)

    def get_action(self, state):
        state = np.reshape(state, [1, args.time_steps,
                                   self.state_dim])
        self.epsilon *= args.eps_decay
        self.epsilon = max(self.epsilon, args.eps_min)
        q_value = self.predict(state)[0]
        if np.random.random() < self.epsilon:
            return random.randint(0, self.action_dim - 1)
        return np.argmax(q_value)

    def train(self, states, targets):
        targets = tf.stop_gradient(targets)
        with tf.GradientTape() as tape:
            logits = self.model(states, training=True)
```

```
        assert targets.shape == logits.shape
        loss = self.compute_loss(targets, logits)
    grads = tape.gradient(loss,
                    self.model.trainable_variables)
    self.opt.apply_gradients(zip(grads,
                    self.model.trainable_variables))
```

6. With the other components implemented, we can begin implementing our `Agent` class:

```
class Agent:
    def __init__(self, env):
        self.env = env
        self.state_dim = \
            self.env.observation_space.shape[0]
        self.action_dim = self.env.action_space.n

        self.states = np.zeros([args.time_steps,
                        self.state_dim])

        self.model = DRQN(self.state_dim,
                        self.action_dim)
        self.target_model = DRQN(self.state_dim,
                        self.action_dim)
        self.update_target()

        self.buffer = ReplayBuffer()

    def update_target(self):
        weights = self.model.model.get_weights()
        self.target_model.model.set_weights(weights)
```

7. In addition to the `train` method in the DRQN class that we implemented in step 6, the crux of the deep recurrent Q-learning algorithm is the q-learning update and experience replay. Let's implement that next:

```
    def replay_experience(self):
        for _ in range(10):
```

```
            states, actions, rewards, next_states, done=\
                self.buffer.sample()
            targets = self.target_model.predict(states)
            next_q_values = self.target_model.\
                        predict(next_states).max(axis=1)
            targets[range(args.batch_size), actions] = (
                rewards + (1 - done) * next_q_values * \
                args.gamma
            )
            self.model.train(states, targets)
```

8. Since the DRQN agent uses recurrent states, let's implement the update_states method to update the recurrent state of the agent:

```
def update_states(self, next_state):
    self.states = np.roll(self.states, -1, axis=0)
    self.states[-1] = next_state
```

9. The next crucial step is to implement the train function to train the agent:

```
def train(self, max_episodes=1000):
    with writer.as_default():
        for ep in range(max_episodes):
            done, episode_reward = False, 0
            self.states = np.zeros([args.time_steps,
                                self.state_dim])
            self.update_states(self.env.reset())
            while not done:
                action = self.model.get_action(
                                self.states)
                next_state, reward, done, _ = \
                                self.env.step(action)
                prev_states = self.states
                self.update_states(next_state)
                self.buffer.store(
                    prev_states, action, reward * \
                    0.01, self.states, done
                )
```

```
                            episode_reward += reward

                    if self.buffer.size() >= args.batch_size:
                        self.replay_experience()
                    self.update_target()
                    print(f"Episode#{ep} \
                        Reward:{episode_reward}")
                    tf.summary.scalar("episode_reward",
        episode_reward, step=ep)
```

10. Finally, let's create the main training loop for the agent:

```
if __name__ == "__main__":
    env = gym.make("Pong-v0")
    agent = Agent(env)
    agent.train(max_episodes=20000)
```

11. To train the DRQN agent in the default environment (CartPole-v0), execute the following command:

```
python ch3-deep-rl-agents/4_drqn.py
```

12. You can also train the DQN agent in any OpenAI Gym-compatible discrete action-space environment using the command-line arguments:

```
python ch3-deep-rl-agents/4_drqn.py –env "MountainCar-v0"
```

How it works...

The DRQN agent uses an LSTM layer, which adds a recurrent learning capability to the agent. The LSTM layer is added to the agent's network in step 5 of the recipe. The other steps in the recipe have similar components as the DQN agent.

Implementing the Asynchronous Advantage Actor-Critic algorithm and A3C agent

The A3C algorithm builds upon the Actor-Critic class of algorithms by using a neural network to approximate the actor (and critic). The actor learns the policy function using a deep neural network, while the critic estimates the value function. The asynchronous nature of the algorithm allows the agent to learn from different parts of the state space, allowing parallel learning and faster convergence. Unlike DQN agents, which use an experience replay memory, the A3C agent uses multiple workers to gather more samples for learning. By the end of this recipe, you will have a complete script to train an A3C agent for any continuous action valued environment of your choice!

Getting ready

To complete this recipe, you will first need to activate the `tf2rl-cookbook` Conda Python virtual environment and `pip install -r requirements.txt`. If the following import statements run without issues, you are ready to get started!

```
import argparse
import os
from datetime import datetime
from multiprocessing import cpu_count
from threading import Thread

import gym
import numpy as np
import tensorflow as tf
from tensorflow.keras.layers import Input, Dense, Lambda
```

Now we can begin.

How to do it...

We will implement an **Asynchronous, Advantage Actor-Critic (A3C)** algorithm by making use of Python's multiprocessing and multithreading capabilities. The following steps will help you to implement a complete A3C agent from scratch using TensorFlow 2.x:

1. First, let's create an argument parser to handle configuration inputs to the script:

```python
parser = argparse.ArgumentParser(prog="TFRL-Cookbook-Ch3-A3C")

parser.add_argument("--env",
default="MountainCarContinuous-v0")

parser.add_argument("--actor-lr", type=float,
default=0.001)

parser.add_argument("--critic-lr", type=float,
default=0.002)

parser.add_argument("--update-interval", type=int,
default=5)

parser.add_argument("--gamma", type=float, default=0.99)

parser.add_argument("--logdir", default="logs")


args = parser.parse_args()
```

2. Let's now create a Tensorboard logger to log useful statistics during the agent's training:

```python
logdir = os.path.join(
    args.logdir, parser.prog, args.env, \
      datetime.now().strftime("%Y%m%d-%H%M%S")
)
print(f"Saving training logs to:{logdir}")
writer = tf.summary.create_file_writer(logdir)
```

3. To have a count of the global episode number, let's define a global variable:

```python
GLOBAL_EPISODE_NUM = 0
```

4. We can now focus on implementing the `Actor` class, which will contain a neural network-based policy to act in the environments:

```
class Actor:
    def __init__(self, state_dim, action_dim,
    action_bound, std_bound):
        self.state_dim = state_dim
        self.action_dim = action_dim
        self.action_bound = action_bound
        self.std_bound = std_bound
        self.model = self.nn_model()
        self.opt = tf.keras.optimizers.Adam(
                                    args.actor_lr)
        self.entropy_beta = 0.01

    def nn_model(self):
        state_input = Input((self.state_dim,))
        dense_1 = Dense(32, activation="relu")\
                        (state_input)
        dense_2 = Dense(32, activation="relu")(dense_1)
        out_mu = Dense(self.action_dim, \
                    activation="tanh")(dense_2)
        mu_output = Lambda(lambda x: x * \
                        self.action_bound)(out_mu)
        std_output = Dense(self.action_dim,
                        activation="softplus")(dense_2)
        return tf.keras.models.Model(state_input,
                            [mu_output, std_output])
```

5. To get an action from the actor given a state, let's define the `get_action` method:

```
    def get_action(self, state):
        state = np.reshape(state, [1, self.state_dim])
        mu, std = self.model.predict(state)
        mu, std = mu[0], std[0]
        return np.random.normal(mu, std,
                            size=self.action_dim)
```

6. Next, to compute the loss, we need to calculate the log of the policy (probability) density function:

```python
def log_pdf(self, mu, std, action):
    std = tf.clip_by_value(std, self.std_bound[0],
                            self.std_bound[1])
    var = std ** 2
    log_policy_pdf = -0.5 * (action - mu) ** 2 / var\
                    - 0.5 * tf.math.log(
        var * 2 * np.pi
    )
    return tf.reduce_sum(log_policy_pdf, 1,
                        keepdims=True)
```

7. Let's now use the `log_pdf` method to compute the actor loss:

```python
def compute_loss(self, mu, std, actions, advantages):
    log_policy_pdf = self.log_pdf(mu, std, actions)
    loss_policy = log_policy_pdf * advantages
    return tf.reduce_sum(-loss_policy)
```

8. As the final step in the `Actor` class implementation, let's define the `train` method:

```python
def train(self, states, actions, advantages):
    with tf.GradientTape() as tape:
        mu, std = self.model(states, training=True)
        loss = self.compute_loss(mu, std, actions,
                                advantages)
    grads = tape.gradient(loss,
                        self.model.trainable_variables)
    self.opt.apply_gradients(zip(grads,
                        self.model.trainable_variables))
    return loss
```

9. With the `Actor` class defined, we can move on to define the `Critic` class:

```python
class Critic:
    def __init__(self, state_dim):
        self.state_dim = state_dim
        self.model = self.nn_model()
```

```
        self.opt = tf.keras.optimizers.Adam\
                        (args.critic_lr)

    def nn_model(self):
        return tf.keras.Sequential(
            [
                Input((self.state_dim,)),
                Dense(32, activation="relu"),
                Dense(32, activation="relu"),
                Dense(16, activation="relu"),
                Dense(1, activation="linear"),
            ]
        )
```

10. Next, let's define the `train` method and a `compute_loss` method to train the critic:

```
    def compute_loss(self, v_pred, td_targets):
        mse = tf.keras.losses.MeanSquaredError()
        return mse(td_targets, v_pred)

    def train(self, states, td_targets):
        with tf.GradientTape() as tape:
            v_pred = self.model(states, training=True)
            assert v_pred.shape == td_targets.shape
            loss = self.compute_loss(v_pred, \
                        tf.stop_gradient(td_targets))
        grads = tape.gradient(loss, \
                        self.model.trainable_variables)
        self.opt.apply_gradients(zip(grads,
                        self.model.trainable_variables))
        return loss
```

11. It is time to implement the `A3CWorker` class based on Python's Thread interface:

```
class A3CWorker(Thread):
    def __init__(self, env, global_actor, global_critic,
    max_episodes):
```

```
Thread.__init__(self)
self.env = env
self.state_dim = \
    self.env.observation_space.shape[0]
self.action_dim = self.env.action_space.shape[0]
self.action_bound = self.env.action_space.high[0]
self.std_bound = [1e-2, 1.0]
self.max_episodes = max_episodes
self.global_actor = global_actor
self.global_critic = global_critic
self.actor = Actor(
    self.state_dim, self.action_dim,
    self.action_bound, self.std_bound
)
self.critic = Critic(self.state_dim)
self.actor.model.set_weights(
    self.global_actor.model.get_weights())
self.critic.model.set_weights(
    self.global_critic.model.get_weights())
```

12. We will be using **n-step Temporal Difference (TD)** learning updates. Therefore, let's define a method to calculate the n-step TD target:

```
def n_step_td_target(self, rewards, next_v_value,
done):
    td_targets = np.zeros_like(rewards)
    cumulative = 0
    if not done:
        cumulative = next_v_value
    for k in reversed(range(0, len(rewards))):
        cumulative = args.gamma * cumulative + \
                        rewards[k]
        td_targets[k] = cumulative
    return td_targets
```

13. We will also need to calculate the advantage values. The advantage value in its simplest form is easy to implement:

```
def advantage(self, td_targets, baselines):
    return td_targets - baselines
```

14. We will split the implementation of the `train` method into the following two steps. First, let's implement the outer loop:

```
def train(self):
    global GLOBAL_EPISODE_NUM
    while self.max_episodes >= GLOBAL_EPISODE_NUM:
        state_batch = []
        action_batch = []
        reward_batch = []
        episode_reward, done = 0, False

        state = self.env.reset()

        while not done:
            # self.env.render()
            action = self.actor.get_action(state)
            action = np.clip(action,
                             -self.action_bound,
                             self.action_bound)
            next_state, reward, done, _ = \
                self.env.step(action)

            state = np.reshape(state, [1,
                                       self.state_dim])
            action = np.reshape(action, [1, 1])
            next_state = np.reshape(next_state,
                                    [1, self.state_dim])
            reward = np.reshape(reward, [1, 1])
            state_batch.append(state)
            action_batch.append(action)
            reward_batch.append(reward)
```

15. In this step, we will complete the `train` method implementation:

```
if len(state_batch) >= args.update_\
interval or done:
    states = np.array([state.squeeze() \
                for state in state_batch])
    actions = np.array([action.squeeze()\
                for action in action_batch])
    rewards = np.array([reward.squeeze()\
                for reward in reward_batch])
    next_v_value = self.critic.model.\
                        predict(next_state)
    td_targets = self.n_step_td_target(
        (rewards + 8) / 8, next_v_value,
        done
    )
    advantages = td_targets - \
        self.critic.model.predict(states)

    actor_loss = self.global_actor.train(
            states, actions, advantages)
    critic_loss = self.global_critic.\
            train(states, td_targets)

    self.actor.model.set_weights(self.\
        global_actor.model.get_weights())
    self.critic.model.set_weights(
        self.global_critic.model.\
        get_weights()
    )

    state_batch = []
    action_batch = []
    reward_batch = []

episode_reward += reward[0][0]
state = next_state[0]
```

```
                    print(f"Episode#{GLOBAL_EPISODE_NUM}\
                        Reward:{episode_reward}")
                tf.summary.scalar("episode_reward",
                                episode_reward,
                                step=GLOBAL_EPISODE_NUM)
            GLOBAL_EPISODE_NUM += 1
```

16. The run method for the A3CWorker thread will simply be the following:

```
        def run(self):
            self.train()
```

17. Next, let's implement the Agent class:

```
    class Agent:
        def __init__(self, env_name,
                        num_workers=cpu_count()):
            env = gym.make(env_name)
            self.env_name = env_name
            self.state_dim = env.observation_space.shape[0]
            self.action_dim = env.action_space.shape[0]
            self.action_bound = env.action_space.high[0]
            self.std_bound = [1e-2, 1.0]

            self.global_actor = Actor(
                self.state_dim, self.action_dim,
                self.action_bound, self.std_bound
            )
            self.global_critic = Critic(self.state_dim)
            self.num_workers = num_workers
```

18. The A3C agent makes use of several concurrent workers. In order to update each of the workers to update the A3C agent, the following code is necessary:

```
    def train(self, max_episodes=20000):
        workers = []

        for i in range(self.num_workers):
```

```
            env = gym.make(self.env_name)
            workers.append(
                A3CWorker(env, self.global_actor,
                        self.global_critic, max_episodes)
            )

        for worker in workers:
            worker.start()

        for worker in workers:
            worker.join()
```

19. With that, our A3C agent implementation is complete, and we are ready to define our main function:

```
if __name__ == "__main__":
    env_name = "MountainCarContinuous-v0"
    agent = Agent(env_name, args.num_workers)
    agent.train(max_episodes=20000)
```

How it works...

In simple terms, the crux of the A3C algorithm can be summarized in the following sequence of steps for each iteration:

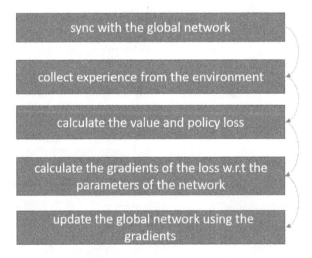

Figure 3.2 – Updating steps in the A3C agent learning iteration

The steps repeat again from top to bottom for the next iteration and so on until convergence.

Implementing the Proximal Policy Optimization algorithm and PPO agent

The **Proximal Policy Optimization (PPO)** algorithm builds upon the work of **Trust Region Policy Optimization (TRPO)** to constrain the new policy to be within a trust region from the old policy. PPO simplifies the implementation of this core idea by using a clipped surrogate objective function that is easier to implement, yet quite powerful and efficient. It is one of the most widely used RL algorithms, especially for continuous control problems. By the end of this recipe, you will have built a PPO agent that you can train in your RL environment of choice.

Getting ready

To complete this recipe, you will first need to activate the tf2rl-cookbook Conda Python virtual environment and pip install -r requirements.txt. If the following import statements run without issues, you are ready to get started!

```
import argparse
import os
from datetime import datetime

import gym
import numpy as np
import tensorflow as tf
from tensorflow.keras.layers import Dense, Input, Lambda
```

We are ready to get started.

How to do it...

The following steps will help you to implement a complete PPO agent from scratch using TensorFlow 2.x:

1. First, let's create an argument parser to handle configuration inputs to the script:

    ```
    parser = argparse.ArgumentParser(prog="TFRL-Cookbook-Ch3-
    PPO")
    ```

```
parser.add_argument("--env", default="Pendulum-v0")
parser.add_argument("--update-freq", type=int, default=5)
parser.add_argument("--epochs", type=int, default=3)
parser.add_argument("--actor-lr", type=float,
default=0.0005)
parser.add_argument("--critic-lr", type=float,
default=0.001)
parser.add_argument("--clip-ratio", type=float,
default=0.1)
parser.add_argument("--gae-lambda", type=float,
default=0.95)
parser.add_argument("--gamma", type=float, default=0.99)
parser.add_argument("--logdir", default="logs")


args = parser.parse_args()
```

2. Next, let's create a Tensorboard logger to log useful statistics during the agent's training:

```
logdir = os.path.join(
    args.logdir, parser.prog, args.env,
    datetime.now().strftime("%Y%m%d-%H%M%S")
)
print(f"Saving training logs to:{logdir}")
writer = tf.summary.create_file_writer(logdir)
```

3. We can now focus on implementing the `Actor` class, which will contain a neural network-based policy to act:

```
class Actor:
    def __init__(self, state_dim, action_dim,
    action_bound, std_bound):
        self.state_dim = state_dim
        self.action_dim = action_dim
        self.action_bound = action_bound
        self.std_bound = std_bound
        self.model = self.nn_model()
        self.opt = \
            tf.keras.optimizers.Adam(args.actor_lr)
```

```python
    def nn_model(self):
        state_input = Input((self.state_dim,))
        dense_1 = Dense(32, activation="relu")\
                        (state_input)
        dense_2 = Dense(32, activation="relu")\
                        (dense_1)
        out_mu = Dense(self.action_dim,
                        activation="tanh")(dense_2)
        mu_output = Lambda(lambda x: x * \
                        self.action_bound)(out_mu)
        std_output = Dense(self.action_dim,
                        activation="softplus")(dense_2)
        return tf.keras.models.Model(state_input,
                        [mu_output, std_output])
```

4. To get an action from the actor given a state, let's define the `get_action` method:

```python
    def get_action(self, state):
        state = np.reshape(state, [1, self.state_dim])
        mu, std = self.model.predict(state)
        action = np.random.normal(mu[0], std[0],
                                size=self.action_dim)
        action = np.clip(action, -self.action_bound,
                        self.action_bound)
        log_policy = self.log_pdf(mu, std, action)

        return log_policy, action
```

5. Next, to compute the loss, we need to calculate the log of the policy (probability) density function:

```python
    def log_pdf(self, mu, std, action):
        std = tf.clip_by_value(std, self.std_bound[0],
                                self.std_bound[1])
        var = std ** 2
        log_policy_pdf = -0.5 * (action - mu) ** 2 / var\
                        - 0.5 * tf.math.log(
```

```
                var * 2 * np.pi
        )
        return tf.reduce_sum(log_policy_pdf, 1,
                            keepdims=True)
```

6. Let's now use the `log_pdf` method to compute the actor loss:

```
def compute_loss(self, log_old_policy,
                log_new_policy, actions, gaes):
    ratio = tf.exp(log_new_policy - \
                tf.stop_gradient(log_old_policy))
    gaes = tf.stop_gradient(gaes)
    clipped_ratio = tf.clip_by_value(
        ratio, 1.0 - args.clip_ratio, 1.0 + \
        args.clip_ratio
    )
    surrogate = -tf.minimum(ratio * gaes, \
                        clipped_ratio * gaes)
    return tf.reduce_mean(surrogate)
```

7. As the final step in the `Actor` class implementation, let's define the `train` method:

```
def train(self, log_old_policy, states, actions,
gaes):
    with tf.GradientTape() as tape:
        mu, std = self.model(states, training=True)
        log_new_policy = self.log_pdf(mu, std,
                                    actions)
        loss = self.compute_loss(log_old_policy,
                        log_new_policy, actions, gaes)
    grads = tape.gradient(loss,
                        self.model.trainable_variables)
    self.opt.apply_gradients(zip(grads,
                        self.model.trainable_variables))
    return loss
```

8. With the `Actor` class defined, we can move on to define the `Critic` class:

```python
class Critic:
    def __init__(self, state_dim):
        self.state_dim = state_dim
        self.model = self.nn_model()
        self.opt = tf.keras.optimizers.Adam(
                                    args.critic_lr)

    def nn_model(self):
        return tf.keras.Sequential(
            [
                    Input((self.state_dim,)),
                    Dense(32, activation="relu"),
                    Dense(32, activation="relu"),
                    Dense(16, activation="relu"),
                    Dense(1, activation="linear"),
            ]
        )
```

9. Next, let's define the `train` method and a `compute_loss` method to train the critic:

```python
    def compute_loss(self, v_pred, td_targets):
        mse = tf.keras.losses.MeanSquaredError()
        return mse(td_targets, v_pred)

    def train(self, states, td_targets):
        with tf.GradientTape() as tape:
            v_pred = self.model(states, training=True)
            assert v_pred.shape == td_targets.shape
            loss = self.compute_loss(v_pred,
                            tf.stop_gradient(td_targets))
        grads = tape.gradient(loss,
                        self.model.trainable_variables)
        self.opt.apply_gradients(zip(grads,
                        self.model.trainable_variables))
        return loss
```

10. It is now time to implement the PPO `Agent` class:

```
class Agent:
    def __init__(self, env):
        self.env = env
        self.state_dim = \
            self.env.observation_space.shape[0]
        self.action_dim = self.env.action_space.shape[0]
        self.action_bound = self.env.action_space.high[0]
        self.std_bound = [1e-2, 1.0]

        self.actor_opt = \
            tf.keras.optimizers.Adam(args.actor_lr)
        self.critic_opt = \
            tf.keras.optimizers.Adam(args.critic_lr)
        self.actor = Actor(
            self.state_dim, self.action_dim,
            self.action_bound, self.std_bound
        )
        self.critic = Critic(self.state_dim)
```

11. We will be using the **Generalized Advantage Estimates (GAE)**. Let's implement a method to calculate the GAE target values:

```
    def gae_target(self, rewards, v_values, next_v_value,
    done):
        n_step_targets = np.zeros_like(rewards)
        gae = np.zeros_like(rewards)
        gae_cumulative = 0
        forward_val = 0

        if not done:
            forward_val = next_v_value

        for k in reversed(range(0, len(rewards))):
            delta = rewards[k] + args.gamma * \
                    forward_val - v_values[k]
            gae_cumulative = args.gamma * \
```

```
                     args.gae_lambda * gae_cumulative + delta
             gae[k] = gae_cumulative
             forward_val = v_values[k]
             n_step_targets[k] = gae[k] + v_values[k]
         return gae, n_step_targets
```

12. We will now split the implementation of the `train` method. First, let's implement the outer loop:

```
def train(self, max_episodes=1000):
    with writer.as_default():
        for ep in range(max_episodes):
            state_batch = []
            action_batch = []
            reward_batch = []
            old_policy_batch = []

            episode_reward, done = 0, False

            state = self.env.reset()
```

13. In this step, we will start the inner loop (per episode) implementation and finish it in the next couple of steps:

```
            while not done:
                # self.env.render()
                log_old_policy, action = \
                    self.actor.get_action(state)

                next_state, reward, done, _ = \
                                self.env.step(action)

                state = np.reshape(state, [1,
                                    self.state_dim])
                action = np.reshape(action, [1, 1])
                next_state = np.reshape(next_state,
                                [1, self.state_dim])
                reward = np.reshape(reward, [1, 1])
```

```
                    log_old_policy = \
                        np.reshape(log_old_policy, [1, 1])


                    state_batch.append(state)
                    action_batch.append(action)
                    reward_batch.append((reward + 8) / 8)
                    old_policy_batch.append(log_old_
        policy)
```

14. In this step, we will use the value predictions made by the PPO algorithm to prepare for the policy update process:

```
        if len(state_batch) >= args.update_freq or done:
                    states = np.array([state.\
                            squeeze() for state \
                            in state_batch])
                    actions = np.array(
                        [action.squeeze() for action\
                        in action_batch]
                    )
                    rewards = np.array(
                        [reward.squeeze() for reward\
                        in reward_batch]
                    )
                    old_policies = np.array(
                        [old_pi.squeeze() for old_pi\
                        in old_policy_batch]
                    )


                    v_values = self.critic.model.\
                            predict(states)
                    next_v_value =self.critic.model.\
                            predict(next_state)


                    gaes, td_targets = \
                        self.gae_target(
                            rewards, v_values, \
```

```
                                    next_v_value, done
            )
```

15. In this step, we will implement the PPO algorithm's policy update steps. These happen inside the inner loop whenever enough of an agent's trajectory information is available in the form of sampled experience batches:

```
actor_losses, critic_losses=[],[]
for epoch in range(args.epochs):
    actor_loss =self.actor.train(
        old_policies, states,\
        actions, gaes
    )
    actor_losses.append(
                        actor_loss)
    critic_loss = self.critic.\
        train(states, td_targets)
    critic_losses.append(
                        critic_loss)
    # Plot mean actor & critic losses
    # on every update
    tf.summary.scalar("actor_loss",
        np.mean(actor_losses), step=ep)
    tf.summary.scalar(
        "critic_loss",
        np.mean(critic_losses),
        step=ep
    )
```

16. As the final step of the `train` method, we will reset the intermediate variables and print a summary of the episode reward obtained by the agent:

```
state_batch = []
action_batch = []
reward_batch = []
old_policy_batch = []
episode_reward += reward[0][0]
state = next_state[0]
print(f"Episode#{ep} \
      Reward:{episode_reward}")
tf.summary.scalar("episode_reward", \
                  episode_reward, \
                  step=ep)
```

17. With that, our PPO agent implementation is complete, and we are ready to define our main function to start training!

```
if __name__ == "__main__":
    env_name = "Pendulum-v0"
    env = gym.make(env_name)
    agent = Agent(env)
    agent.train(max_episodes=20000)
```

How it works...

The PPO algorithm uses clipping to form a surrogate loss function, and uses multiple epochs of **Stochastic Gradient Decent/Ascent** (**SGD**) optimization per the policy update. The clipping introduced by PPO reduces the effective change that can be applied to the policy, thereby improving the stability of the policy while learning.

The PPO agent uses actor(s) to collect samples from the environment using the latest policy parameters. The loop defined in step 15 of the recipe samples a mini-batch of experience and trains the network for n epochs (passed as the `--epoch` argument to the script) using the clipped surrogate objective function. The process is then repeated with new samples of experiences.

Implementing the Deep Deterministic Policy Gradient algorithm and DDPG agent

Deterministic Policy Gradient (DPG) is a type of Actor-Critic RL algorithm that uses two neural networks: one for estimating the action value function, and the other for estimating the optimal target policy. The **Deep Deterministic Policy Gradient (DDPG)** agent builds upon the idea of DPG and is quite efficient compared to vanilla Actor-Critic agents due to the use of deterministic action policies. By completing this recipe, you will have access to a powerful agent that can be trained efficiently in a variety of RL environments.

Getting ready

To complete this recipe, you will first need to activate the `tf2rl-cookbook` Conda Python virtual environment and `pip install -r requirements.txt`. If the following import statements run without issues, you are ready to get started!

```
import argparse
import os
import random
from collections import deque
from datetime import datetime

import gym
import numpy as np
import tensorflow as tf
from tensorflow.keras.layers import Dense, Input, Lambda, concatenate
```

Now we can begin.

How to do it...

The following steps will help you to implement a complete DDPG agent from scratch using TensorFlow 2.x:

1. Let's first create an argument parser to handle command-line configuration inputs to the script:

```
parser = argparse.ArgumentParser(prog="TFRL-Cookbook-Ch3-
DDPG")
parser.add_argument("--env", default="Pendulum-v0")
parser.add_argument("--actor_lr", type=float,
default=0.0005)
parser.add_argument("--critic_lr", type=float,
default=0.001)
parser.add_argument("--batch_size", type=int, default=64)
parser.add_argument("--tau", type=float, default=0.05)
parser.add_argument("--gamma", type=float, default=0.99)
parser.add_argument("--train_start", type=int,
default=2000)
parser.add_argument("--logdir", default="logs")

args = parser.parse_args()
```

2. Let's create a Tensorboard logger to log useful statistics during the agent's training:

```
logdir = os.path.join(
    args.logdir, parser.prog, args.env, \
    datetime.now().strftime("%Y%m%d-%H%M%S")
)
print(f"Saving training logs to:{logdir}")
writer = tf.summary.create_file_writer(logdir)
```

3. Let's now implement an experience replay memory:

```
class ReplayBuffer:
    def __init__(self, capacity=10000):
        self.buffer = deque(maxlen=capacity)

    def store(self, state, action, reward, next_state,
            done):
```

```
            self.buffer.append([state, action, reward,
                              next_state, done])

    def sample(self):
        sample = random.sample(self.buffer,
                              args.batch_size)
        states, actions, rewards, next_states, done = \
                          map(np.asarray, zip(*sample))
        states = np.array(states).reshape(
                                    args.batch_size, -1)
        next_states = np.array(next_states).\
                        reshape(args.batch_size, -1)
        return states, actions, rewards, next_states, \
        done

    def size(self):
        return len(self.buffer)
```

4. We can now focus on implementing the `Actor` class, which will contain a neural network-based policy to act:

```
class Actor:
    def __init__(self, state_dim, action_dim,
    action_bound):
        self.state_dim = state_dim
        self.action_dim = action_dim
        self.action_bound = action_bound
        self.model = self.nn_model()
        self.opt = tf.keras.optimizers.Adam(args.actor_
lr)

    def nn_model(self):
        return tf.keras.Sequential(
            [
                Input((self.state_dim,)),
                Dense(32, activation="relu"),
                Dense(32, activation="relu"),
```

```
            Dense(self.action_dim,
                 activation="tanh"),
            Lambda(lambda x: x * self.action_bound),
        ]
    )
```

5. To get an action from the actor given a state, let's define the `get_action` method:

```
def get_action(self, state):
    state = np.reshape(state, [1, self.state_dim])
    return self.model.predict(state)[0]
```

6. Next, we'll implement a predict function to return the predictions made by the actor network:

```
def predict(self, state):
    return self.model.predict(state)
```

7. As the final step in the `Actor` class implementation, let's define the `train` method:

```
def train(self, states, q_grads):
    with tf.GradientTape() as tape:
        grads = tape.gradient(
            self.model(states),
            self.model.trainable_variables, -q_grads
        )
    self.opt.apply_gradients(zip(grads,
                    self.model.trainable_variables))
```

8. With the `Actor` class defined, we can move on to define the `Critic` class:

```
class Critic:
    def __init__(self, state_dim, action_dim):
        self.state_dim = state_dim
        self.action_dim = action_dim
        self.model = self.nn_model()
        self.opt = \
            tf.keras.optimizers.Adam(args.critic_lr)
```

```python
def nn_model(self):
    state_input = Input((self.state_dim,))
    s1 = Dense(64, activation="relu")(state_input)
    s2 = Dense(32, activation="relu")(s1)
    action_input = Input((self.action_dim,))
    a1 = Dense(32, activation="relu")(action_input)
    c1 = concatenate([s2, a1], axis=-1)
    c2 = Dense(16, activation="relu")(c1)
    output = Dense(1, activation="linear")(c2)
    return tf.keras.Model([state_input,
                           action_input], output)
```

9. In this step, we will be implementing a method to calculate the gradients of the Q function:

```python
def q_gradients(self, states, actions):
    actions = tf.convert_to_tensor(actions)
    with tf.GradientTape() as tape:
        tape.watch(actions)
        q_values = self.model([states, actions])
        q_values = tf.squeeze(q_values)
    return tape.gradient(q_values, actions)
```

10. As a convenience method, let's also define a `predict` function to return the critic network's prediction:

```python
def predict(self, inputs):
    return self.model.predict(inputs)
```

11. Next, let's define the `train` method and a `compute_loss` method to train the critic:

```python
def train(self, states, actions, td_targets):
    with tf.GradientTape() as tape:
        v_pred = self.model([states, actions],
                            training=True)
        assert v_pred.shape == td_targets.shape
        loss = self.compute_loss(v_pred,
```

```
                               tf.stop_gradient(td_targets))
        grads = tape.gradient(loss,
                         self.model.trainable_variables)
        self.opt.apply_gradients(zip(grads,
                         self.model.trainable_variables))
        return loss
```

12. It is now time to implement the DDPG `Agent` class:

```
class Agent:
    def __init__(self, env):
        self.env = env
        self.state_dim = \
            self.env.observation_space.shape[0]
        self.action_dim = self.env.action_space.shape[0]
        self.action_bound = self.env.action_space.high[0]

        self.buffer = ReplayBuffer()

        self.actor = Actor(self.state_dim, \
                           self.action_dim,
                           self.action_bound)
        self.critic = Critic(self.state_dim,
                             self.action_dim)

        self.target_actor = Actor(self.state_dim,
                                  self.action_dim,
                                  self.action_bound)
        self.target_critic = Critic(self.state_dim,
                                    self.action_dim)

        actor_weights = self.actor.model.get_weights()
        critic_weights = self.critic.model.get_weights()
        self.target_actor.model.set_weights(
                                    actor_weights)
        self.target_critic.model.set_weights(
                                    critic_weights)
```

13. Let's now implement the `update_target` method to update the actor and critic network's weights with that of the respective target networks:

```
def update_target(self):
    actor_weights = self.actor.model.get_weights()
    t_actor_weights = \
        self.target_actor.model.get_weights()
    critic_weights = self.critic.model.get_weights()
    t_critic_weights = \
        self.target_critic.model.get_weights()

    for i in range(len(actor_weights)):
        t_actor_weights[i] = (
            args.tau * actor_weights[i] + \
            (1 - args.tau) * t_actor_weights[i]
        )

    for i in range(len(critic_weights)):
        t_critic_weights[i] = (
            args.tau * critic_weights[i] + \
            (1 - args.tau) * t_critic_weights[i]
        )

    self.target_actor.model.set_weights(
                            t_actor_weights)
    self.target_critic.model.set_weights(
                            t_critic_weights)
```

14. Next, let's implement a helper method to calculate the TD targets:

```
def get_td_target(self, rewards, q_values, dones):
    targets = np.asarray(q_values)
    for i in range(q_values.shape[0]):
        if dones[i]:
            targets[i] = rewards[i]
        else:
            targets[i] = args.gamma * q_values[i]
    return targets
```

15. The purpose of the Deterministic Policy Gradient algorithm is to add noise to the actions sampled from the deterministic policy. Let's use the **Ornstein-Uhlenback (OU)** process to generate noise:

```
def add_ou_noise(self, x, rho=0.15, mu=0, dt=1e-1,
  sigma=0.2, dim=1):
    return (
        x + rho * (mu - x) * dt + sigma * \
        np.sqrt(dt) * np.random.normal(size=dim)
    )
```

16. In this step, we will use experience replay to update the actor and critic:

```
def replay_experience(self):
    for _ in range(10):
        states, actions, rewards, next_states, \
            dones = self.buffer.sample()
        target_q_values = self.target_critic.predict(
            [next_states, self.target_actor.\
            predict(next_states)]
        )
        td_targets = self.get_td_target(rewards,
                            target_q_values, dones)

        self.critic.train(states, actions,
                            td_targets)

        s_actions = self.actor.predict(states)
        s_grads = self.critic.q_gradients(states,
                                        s_actions)
        grads = np.array(s_grads).reshape((-1,
                                self.action_dim))
        self.actor.train(states, grads)
        self.update_target()
```

17. With all the components we have implemented, we are now ready to put them together in the `train` method:

```
def train(self, max_episodes=1000):
    with writer.as_default():
        for ep in range(max_episodes):
            episode_reward, done = 0, False
            state = self.env.reset()
            bg_noise = np.zeros(self.action_dim)
            while not done:
                # self.env.render()
                action = self.actor.get_action(state)
                noise = self.add_ou_noise(bg_noise, \
                                    dim=self.action_dim)
                action = np.clip(
                    action + noise, -self.action_\
                    bound, self.action_bound
                )
                next_state, reward, done, _ = \
                                self.env.step(action)
                self.buffer.store(state, action, \
                    (reward + 8) / 8, next_state, done)
                bg_noise = noise
                episode_reward += reward
                state = next_state
            if (
                self.buffer.size() >= args.batch_size
                and self.buffer.size() >= \
                    args.train_start
            ):
                self.replay_experience()
            print(f"Episode#{ep} \
                    Reward:{episode_reward}")
            tf.summary.scalar("episode_reward",
                            episode_reward, step=ep)
```

18. With that, our DDPG agent implementation is complete, and we are ready to define our main function to start training!

```python
if __name__ == "__main__":
    env_name = "Pendulum-v0"
    env = gym.make(env_name)
    agent = Agent(env)
    agent.train(max_episodes=20000)
```

How it works...

The DDPG agent estimates two quantities – the Q-value function and the optimal policy. DDPG combines the ideas introduced in DQN and DPG. DDPG uses a policy gradient update rule in addition to the ideas introduced in DQN, as can be seen in the update steps defined in step 14.

4
Reinforcement Learning in the Real World – Building Cryptocurrency Trading Agents

Deep reinforcement learning (deep RL) agents have a lot of potential when it comes to solving challenging problems in the real world and a lot of opportunities exist. However, only a few successful stories of using deep RL agents in the real world beyond games exist due to the various challenges associated with real-world deployments of RL agents. This chapter contains recipes that will help you successfully develop RL agents for an interesting and rewarding real-world problem: **cryptocurrency trading**. The recipes in this chapter contain information on how to implement custom OpenAI Gym-compatible learning environments for cryptocurrency trading with both discrete and continuous-value action spaces. In addition, you will learn how to build and train RL agents for trading cryptocurrency. Trading learning environments will also be provided.

Specifically, the following recipes will be covered in this chapter:

- Building a Bitcoin trading RL platform using real market data
- Building an Ethereum trading RL platform using price charts
- Building an advanced cryptocurrency trading platform for RL agents
- Training a cryptocurrency trading bot using RL

Let's get started!

Technical requirements

The code in the book has been extensively tested on Ubuntu 18.04 and Ubuntu 20.04 and should work with later versions of Ubuntu if Python 3.6+ is available. With Python 3.6+ installed, along with the necessary Python packages listed at the start of each of recipe, the code should run fine on Windows and macOS X too. You should create and use a Python virtual environment named `tf2rl-cookbook` to install the packages and run the code in this book. Installing Miniconda or Anaconda for Python virtual environment management is recommended.

The complete code for each recipe in each chapter is available here: `https://github.com/PacktPublishing/Tensorflow-2-Reinforcement-Learning-Cookbook`.

Building a Bitcoin trading RL platform using real market data

This recipe will help you build a cryptocurrency trading RL environment for your agents. This environment simulates a Bitcoin trading exchange based on real-world data from the Gemini cryptocurrency exchange. In this environment, your RL agent can place buy/sell/hold trades and get rewards based on the profit/loss it makes, starting with an initial cash balance in the agent's trading account.

Getting ready

To complete this recipe, make sure you have the latest version. You will need to activate the `tf2rl-cookbook` Python/conda virtual environment. Make sure to update the environment so that it matches the latest conda environment specification file (`tfrl-cookbook.yml`) in this cookbook's code repository. If the following `import` statements run without issues, you are ready to get started:

```
import os
import random
from typing import Dict

import gym
import numpy as np
import pandas as pd
from gym import spaces
```

Now, let's begin!

How to do it...

Follow these steps to learn how to implement `CryptoTradingEnv`:

1. Let's begin by importing the necessary Python modules.
2. We'll also be using the `TradeVisualizer` class implemented in `trading_utils.py`. We'll discuss this in more deail when we actually use it:

    ```
    from trading_utils import TradeVisualizer
    ```

3. To make it easy to configure the cryptocurrency trading environment, we will set up an environment config dictionary. Notice that our cryptocurrency trading environment has been configured so that we can trade Bitcoin based on real data from the Gemini cryptocurrency exchange:

```
env_config = {
    "exchange": "Gemini", # Cryptocurrency exchange
    # (Gemini, coinbase, kraken, etc.)
    "ticker": "BTCUSD", # CryptoFiat
    "frequency": "daily", # daily/hourly/minutes
    "opening_account_balance": 100000,
    # Number of steps (days) of data provided to the
    # agent in one observation.
    "observation_horizon_sequence_length": 30,
    "order_size": 1, # Number of coins to buy per
    # buy/sell order
}
```

4. Let's begin our `CryptoTradingEnv` class definition:

```
class CryptoTradingEnv(gym.Env):
    def __init__(self, env_config: Dict = env_config):
        super(CryptoTradingEnv, self).__init__()
        self.ticker = env_config.get("ticker", "BTCUSD")
        data_dir = os.path.join(os.path.dirname(os.path.\
                        realpath(__file__)), "data")
        self.exchange = env_config["exchange"]
        freq = env_config["frequency"]
        if freq == "daily":
            self.freq_suffix = "d"
        elif freq == "hourly":
            self.freq_suffix = "1hr"
        elif freq == "minutes":
            self.freq_suffix = "1min"
```

5. We'll be using a file object as our cryptocurrency exchange data source. We must make sure that the data source exists before loading/streaming the data into memory:

```
self.ticker_file_stream = os.path.join(
    f"{data_dir}",
    f"{'_'.join([self.exchange, self.ticker,
               self.freq_suffix])}.csv",
)
assert os.path.isfile(
    self.ticker_file_stream
), f"Cryptocurrency data file stream not found \
    at: data/{self.ticker_file_stream}.csv"
# Cryptocurrency exchange data stream. An offline
# file stream is used. Alternatively, a web
# API can be used to pull live data.
self.ohlcv_df = pd.read_csv(self.ticker_file_\
    stream, skiprows=1).sort_values(by="Date"
)
```

6. The opening balance in the Agent's account is configured using `env_config`. Let's initialize the opening account balance based on the configured value:

```
self.opening_account_balance = env_config["opening_
account_balance"]
```

7. Next, let's define the action and observation space for this cryptocurrency trading environment using the standard space type definitions provided by the OpenAI Gym library:

```
# Action: 0-> Hold; 1-> Buy; 2 ->Sell;
self.action_space = spaces.Discrete(3)

self.observation_features = [
    "Open",
    "High",
    "Low",
    "Close",
```

```
        "Volume BTC",
        "Volume USD",
    ]
    self.horizon = env_config.get(
            "observation_horizon_sequence_length")
    self.observation_space = spaces.Box(
        low=0,
        high=1,
        shape=(len(self.observation_features),
                self.horizon + 1),
        dtype=np.float,
    )
```

8. Let's define the trade order size that will be executed when the agent places a trade:

```
    self.order_size = env_config.get("order_size")
```

9. With that, we have successfully initialized the environment! Now, let's move on and define the step (...) method. You will notice that we have simplified the implementation of the step (...) method for ease of understanding using two helper member methods: self.execute_trade_action and self.get_observation. We'll define these helper member methods later, once we have finished implementing the basic RL Gym environment methods (step, reset, and render). Now, let's look at the implementation of the step method:

```
def step(self, action):
    # Execute one step within the trading environment
    self.execute_trade_action(action)

    self.current_step += 1

    reward = self.account_value - \
            self.opening_account_balance
        # Profit (loss)
    done = self.account_value <= 0 or \
        self.current_step >= len(
        self.ohlcv_df.loc[:, "Open"].values
```

```
        )

        obs = self.get_observation()

        return obs, reward, done, {}
```

10. Now, let's define the `reset()` method, which will be executed at the start of every episode:

```
def reset(self):
    # Reset the state of the environment to an
    # initial state
    self.cash_balance = self.opening_account_balance
    self.account_value = self.opening_account_balance
    self.num_coins_held = 0
    self.cost_basis = 0
    self.current_step = 0
    self.trades = []
    if self.viz is None:
        self.viz = TradeVisualizer(
            self.ticker,
            self.ticker_file_stream,
            "TFRL-Cookbook Ch4-CryptoTradingEnv",
            skiprows=1,  # Skip the first line with
            # the data download source URL
        )
    return self.get_observation()
```

11. As the next step, we'll define the `render()` method, which will provide us with a view into the cryptocurrency trading environment so that we understand what's going on! This is where we will be using the `TradeVisualizer` class from the `trading_utils.py` file. `TradeVisualizer` helps us visualize the live account balance of the Agent as the Agent learns in the environment. The visualizer also provides a visual indication of the buy and sell trades that the Agent performs by performing actions in the environment. A sample screenshot of the output from the `render()` method has been provided here for your reference:

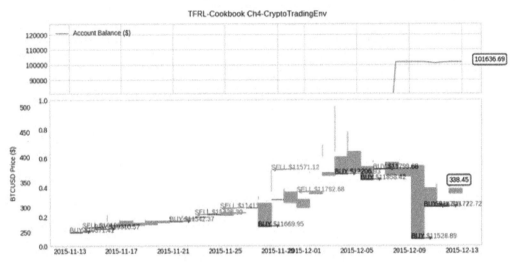

Figure 4.1 – A sample rendering of the CryptoTradingEnv environment

Now, let's implement the `render()` method:

```
def render(self, **kwargs):
    # Render the environment to the screen

    if self.current_step > self.horizon:
        self.viz.render(
            self.current_step,
            self.account_value,
            self.trades,
            window_size=self.horizon,
        )
```

12. Next, we'll implement a method that will close all the visualization windows once the training is complete:

```
def close(self):
    if self.viz is not None:
        self.viz.close()
        self.viz = None
```

13. Now, we can implement the `execute_trade_action` method, which we used in the `step` (...) method earlier in Step 9. We'll split the implementation into three steps, one for each order type: Hold, Buy, and Sell. Let's start with the Hold order type as that's the simplest. You will see why in a bit!

```
def execute_trade_action(self, action):
    if action == 0:  # Hold position
        return
```

14. We actually need to implement one more intermediate step before we can move on and implement the Buy and Sell order execution logic. Here, we must determine the order type (buy versus sell) and then the price of the Bitcoin at the current simulated time:

```
order_type = "buy" if action == 1 else "sell"


# Stochastically determine the current price
# based on Market Open & Close
current_price = random.uniform(
    self.ohlcv_df.loc[self.current_step, "Open"],
    self.ohlcv_df.loc[self.current_step,
                      "Close"],
)
```

15. Now, we are ready to implement the logic for executing a Buy trade order, as follows:

```python
if order_type == "buy":
    allowable_coins = \
        int(self.cash_balance / current_price)
    if allowable_coins < self.order_size:
        # Not enough cash to execute a buy order
        return
    # Simulate a BUY order and execute it at
    # current_price
    num_coins_bought = self.order_size
    current_cost = self.cost_basis * \
                    self.num_coins_held
    additional_cost = num_coins_bought * \
                    current_price

    self.cash_balance -= additional_cost
    self.cost_basis = (current_cost + \
        additional_cost) / (
        self.num_coins_held + num_coins_bought
    )
    self.num_coins_held += num_coins_bought
```

16. Let's update the `trades` list with the latest buy trade:

```python
self.trades.append(
    {
        "type": "buy",
        "step": self.current_step,
        "shares": num_coins_bought,
        "proceeds": additional_cost,
    }
)
```

17. The next step is to implement the logic for executing Sell trade orders:

```python
elif order_type == "sell":
    # Simulate a SELL order and execute it at
    # current_price
    if self.num_coins_held < self.order_size:
        # Not enough coins to execute a sell
        # order
        return
    num_coins_sold = self.order_size
    self.cash_balance += num_coins_sold * \
                        current_price
    self.num_coins_held -= num_coins_sold
    sale_proceeds = num_coins_sold * \
                    current_price

    self.trades.append(
        {
            "type": "sell",
            "step": self.current_step,
            "shares": num_coins_sold,
            "proceeds": sale_proceeds,
        }
    )
```

18. To finish up our trade execution function, we need to add a couple of lines of code that will update the account value once the trade order has been executed:

```python
if self.num_coins_held == 0:
    self.cost_basis = 0
# Update account value
self.account_value = self.cash_balance + \
                    self.num_coins_held * \
                    current_price
```

19. With that, we have finished implementing a Bitcoin trading RL environment powered by real BTCUSD data from the Gemini cryptocurrency exchange! Let's look at how we can easily create the environment and run a sample, rather than using a random agent in this environment with just six lines of code:

```
if __name__ == "__main__":
    env = CryptoTradingEnv()
    obs = env.reset()
    for _ in range(600):
        action = env.action_space.sample()
        next_obs, reward, done, _ = env.step(action)
        env.render()
```

You should see the sample random agent acting in the `CryptoTradingEnv` environment. The `env.render()` function should produce a rendering that looks similar to the following:

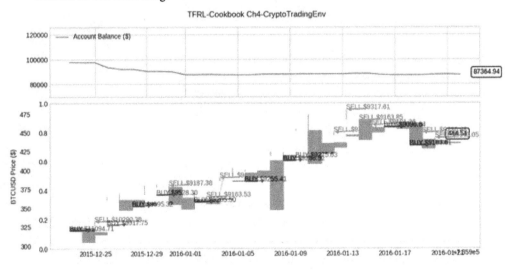

Figure 4.2 – A rendering of the CryptoTradingEnv environment showing the agent's current account balance and the buy/sell trade being executed

Now, let's see how this all works.

How it works...

In this recipe, we implemented the `CryptoTradingEnv` function, which offers tabular observations of shape (6, horizon + 1), where the horizon can be configured through the `env_config` dictionary. The horizon parameter specifies the horizon of the duration of the time window (for example, 3 days) that the Agent is allowed to observe the cryptocurrency market data at every step before making a trade. Once the Agent takes one of the allowed discrete actions – 0(hold), 1(buy), or 2(sell) – the appropriate trade is executed at the current exchange price of the cryptocurrency (Bitcoin) and the trading account balance is updated accordingly. The Agent will also receive a reward based on the profit (or loss) that's made through the trades from the start of the episode.

Building an Ethereum trading RL platform using price charts

This recipe will teach you to implement an Ethereum cryptocurrency trading environment for RL Agents with visual observations. The Agent will observe a price chart with Open, High, Low, Close, and Volume information over a specified time period to take an action (Hold, Buy, or Sell). The objective of the Agent is to maximize its reward, which is the profit you would make if you deployed the Agent to trade in your account!

Getting ready

To complete this recipe, make sure you have the latest version. You will need to activate the `tf2rl-cookbook` Python/conda virtual environment. Make sure that will update the environment so that it matches the latest conda environment specification file (`tfrl-cookbook.yml`), which can be found in this cookbook's code repository. If the following `import` statements run without any issues, you are ready to get started:

```
import os
import random
from typing import Dict


import cv2
import gym
import numpy as np
import pandas as pd
from gym import spaces
from trading_utils import TradeVisualizer
```

How to do it...

Let's follow the OpenAI Gym framework in order to implement our learning environment interface. We will add some logic that will simulate cryptocurrency trade execution and reward the agent appropriately since this will aid your learning.

Follow these steps to complete your implementation:

1. Let's begin by configuring the environment using a dictionary:

```
env_config = {
    "exchange": "Gemini",  # Cryptocurrency exchange
    # (Gemini, coinbase, kraken, etc.)
    "ticker": "ETHUSD",  # CryptoFiat
    "frequency": "daily",  # daily/hourly/minutes
    "opening_account_balance": 100000,
    # Number of steps (days) of data provided to the
    # agent in one observation
    "observation_horizon_sequence_length": 30,
    "order_size": 1,  # Number of coins to buy per
    # buy/sell order
}
```

2. Let's define the `CryptoTradingVisualEnv` class and load the settings from `env_config`:

```
class CryptoTradingVisualEnv(gym.Env):
    def __init__(self, env_config: Dict = env_config):
        """Cryptocurrency trading environment for RL
        agents
        The observations are cryptocurrency price info
        (OHLCV) over a horizon as specified in
        env_config. Action space is discrete to perform
        buy/sell/hold trades.
        Args:
            ticker(str, optional): Ticker symbol for the\
            crypto-fiat currency pair.
            Defaults to "ETHUSD".
            env_config (Dict): Env configuration values
```

```
    """
    super(CryptoTradingVisualEnv, self).__init__()
    self.ticker = env_config.get("ticker", "ETHUSD")
    data_dir = os.path.join(os.path.dirname(os.path.\
                        realpath(__file__)), "data")
    self.exchange = env_config["exchange"]
    freq = env_config["frequency"]
```

3. As the next step, based on the frequency configuration for the market data feed, let's load the cryptocurrency exchange data from the input stream:

```
if freq == "daily":
        self.freq_suffix = "d"
elif freq == "hourly":
        self.freq_suffix = "1hr"
elif freq == "minutes":
        self.freq_suffix = "1min"

self.ticker_file_stream = os.path.join(
        f"{data_dir}",
        f"{'_'.join([self.exchange, self.ticker, \
                    self.freq_suffix])}.csv",
)
assert os.path.isfile(
        self.ticker_file_stream
), f"Cryptocurrency exchange data file stream \
not found at: data/{self.ticker_file_stream}.csv"
# Cryptocurrency exchange data stream. An offline
# file stream is used. Alternatively, a web
# API can be used to pull live data.
self.ohlcv_df = pd.read_csv(self.ticker_file_\
        stream, skiprows=1).sort_values(
        by="Date"
)
```

4. Let's initialize other environment class variables and define the state and
    action space:

```
self.opening_account_balance = \
    env_config["opening_account_balance"]
# Action: 0-> Hold; 1-> Buy; 2 ->Sell;
self.action_space = spaces.Discrete(3)

self.observation_features = [
    "Open",
    "High",
    "Low",
    "Close",
    "Volume ETH",
    "Volume USD",
]
self.obs_width, self.obs_height = 128, 128
self.horizon = env_config.get("
    observation_horizon_sequence_length")
self.observation_space = spaces.Box(
    low=0, high=255, shape=(128, 128, 3),
    dtype=np.uint8,
)
self.order_size = env_config.get("order_size")
self.viz = None   # Visualizer
```

5. Let's define the `reset` method in order to (re)initialize the environment class
    variables:

```
def reset(self):
    # Reset the state of the environment to an
    # initial state
    self.cash_balance = self.opening_account_balance
    self.account_value = self.opening_account_balance
    self.num_coins_held = 0
    self.cost_basis = 0
    self.current_step = 0
```

```
        self.trades = []
        if self.viz is None:
            self.viz = TradeVisualizer(
                self.ticker,
                self.ticker_file_stream,
                "TFRL-Cookbook\
                    Ch4-CryptoTradingVisualEnv",
                skiprows=1,
            )

        return self.get_observation()
```

6. The key feature of this environment is that the Agent's observations are images of the price chart, similar to the one you can see on a human trader's computer screen. This chart contains flashy plots with red and green bars and candles! Let's define the get_observation method in order to return an image of the charting screen:

```
def get_observation(self):
    """Return a view of the Ticker price chart as
        image observation

    Returns:
        img_observation(np.ndarray): Image of ticker
        candle stick plot with volume bars as
        observation
    """
    img_observation = \
        self.viz.render_image_observation(
        self.current_step, self.horizon
    )
    img_observation = cv2.resize(
        img_observation, dsize=(128, 128),
        interpolation=cv2.INTER_CUBIC
    )

    return img_observation
```

7. Now, we'll implement the trade execution logic of the trading environment. The current price of the Ethereum cryptocurrency (in USD) must be extracted from the market data stream (a file, in this case):

```python
def execute_trade_action(self, action):
    if action == 0:  # Hold position
        return
    order_type = "buy" if action == 1 else "sell"

    # Stochastically determine the current price
    # based on Market Open & Close
    current_price = random.uniform(
        self.ohlcv_df.loc[self.current_step, "Open"],
        self.ohlcv_df.loc[self.current_step,
                          "Close"],
    )
```

8. If the Agent decides to execute a buy order, we must calculate the number of Ethereum tokens/coins the Agent can buy in a single step and execute the "Buy" order at the simulated exchange:

```python
    # Buy Order
    allowable_coins = \
        int(self.cash_balance / current_price)
    if allowable_coins < self.order_size:
        # Not enough cash to execute a buy order
        return
    # Simulate a BUY order and execute it at
    # current_price
    num_coins_bought = self.order_size
    current_cost = self.cost_basis * \
                   self.num_coins_held
    additional_cost = num_coins_bought * \
                      current_price

    self.cash_balance -= additional_cost
    self.cost_basis = \
```

```
                    (current_cost + additional_cost) / (
                    self.num_coins_held + num_coins_bought
        )
        self.num_coins_held += num_coins_bought

        self.trades.append(
            {
                "type": "buy",
                "step": self.current_step,
                "shares": num_coins_bought,
                "proceeds": additional_cost,
            }
        )
```

9. Instead, if the Agent decides to sell, the following logic will execute the sell order:

```
        # Simulate a SELL order and execute it at
        # current_price
        if self.num_coins_held < self.order_size:
            # Not enough coins to execute a sell
            # order
            return
        num_coins_sold = self.order_size
        self.cash_balance += num_coins_sold * \
                             current_price
        self.num_coins_held -= num_coins_sold
        sale_proceeds = num_coins_sold * \
                        current_price

        self.trades.append(
            {
                "type": "sell",
                "step": self.current_step,
                "shares": num_coins_sold,
                "proceeds": sale_proceeds,
            }
        )
```

10. Let's update the account balance to reflect the effect of the Buy/Sell trade:

```python
if self.num_coins_held == 0:
    self.cost_basis = 0
# Update account value
self.account_value = self.cash_balance + \
                        self.num_coins_held * \
                        current_price
```

11. We are now ready to implement the `step` method:

```python
def step(self, action):
    # Execute one step within the trading environment
    self.execute_trade_action(action)

    self.current_step += 1

    reward = self.account_value - \
        self.opening_account_balance  # Profit (loss)
    done = self.account_value <= 0 or \
            self.current_step >= len(
        self.ohlcv_df.loc[:, "Open"].values
    )

    obs = self.get_observation()

    return obs, reward, done, {}
```

12. Let's implement a method that will render the current state as an image to the screen. This will help us understand what's going on in the environment while the Agent is learning to trade:

```python
def render(self, **kwargs):
    # Render the environment to the screen

    if self.current_step > self.horizon:
        self.viz.render(
            self.current_step,
```

```
                self.account_value,
                self.trades,
                window_size=self.horizon,
        )
```

13. That completes our implementation! Let's quickly check out the environment by using a random agent:

```
if __name__ == "__main__":
    env = CryptoTradingVisualEnv()
    obs = env.reset()
    for _ in range(600):
        action = env.action_space.sample()
        next_obs, reward, done, _ = env.step(action)
        env.render()
```

You should see the sample random agent acting in `CryptoTradinVisualEnv`, wherein the agent receives visual/image observations similar to the one shown here:

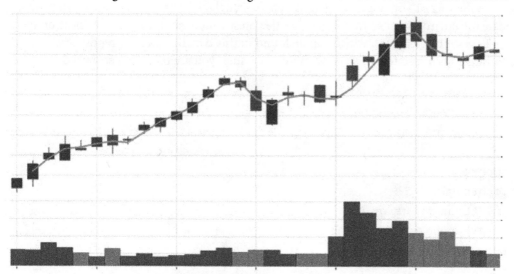

Figure 4.3 – Sample observation sent to the learning Agent

That's it for this recipe!

How it works...

In this recipe, we implemented a visual Ethereum cryptocurrency trading environment that provides images as input to the agents. The images contain charting information, such as Open, High, Low, Close, and Volume data. This chart looks like what a human trader's screen will look like and informs the agent about the current market signals.

Building an advanced cryptocurrency trading platform for RL agents

Instead of allowing the Agent to only take discrete actions, such as buying/selling/holding a pre-set amount of Bitcoin or Ethereum tokens, what if we allowed the Agent to decide how many crypto coins/tokens it would like to buy or sell? That is exactly what this recipe will allow you to create in the form of a `CryptoTradingVisualContinuousEnv` RL environment.

Getting ready

To complete this recipe, you need to ensure you have the latest version. You will need to activate the `tf2rl-cookbook` Python/conda virtual environment. Make sure that you update the environment so that it matches the latest conda environment specification file (`tfrl-cookbook.yml`), which can be found in this cookbook's code repository. If the following `import` statements run without any issues, you are ready to get started:

```
import os
import random
from typing import Dict


import cv2
import gym
import numpy as np
import pandas as pd
from gym import spaces


from trading_utils import TradeVisualizer
```

How to do it...

This is going to be a complex environment as it uses high-dimensional images as observations and allows for continuous, real-value actions to be performed. However, you are likely familiar with the components of this recipe due to having experience implementing the previous recipes in this chapter.

Let's get started:

1. First, we must define the configuration parameters that are allowed for this environment:

```
env_config = {
    "exchange": "Gemini",  # Cryptocurrency exchange
    # (Gemini, coinbase, kraken, etc.)
    "ticker": "BTCUSD",  # CryptoFiat
    "frequency": "daily",  # daily/hourly/minutes
    "opening_account_balance": 100000,
    # Number of steps (days) of data provided to the
    # agent in one observation
    "observation_horizon_sequence_length": 30,
}
```

2. Let's jump right into the definition of the learning environment class:

```
class CryptoTradingVisualContinuousEnv(gym.Env):
    def __init__(self, env_config: Dict = env_config):
        """Cryptocurrency trading environment for RL
        agents with continuous action space

        Args:
            ticker (str, optional): Ticker symbol for the
            crypto-fiat currency pair.
            Defaults to "BTCUSD".
            env_config (Dict): Env configuration values
        """
        super(CryptoTradingVisualContinuousEnv,
            self).__init__()
        self.ticker = env_config.get("ticker", "BTCUSD")
```

```
data_dir = os.path.join(os.path.dirname(os.path.\
                    realpath(__file__)), "data")
self.exchange = env_config["exchange"]
freq = env_config["frequency"]
if freq == "daily":
    self.freq_suffix = "d"
elif freq == "hourly":
    self.freq_suffix = "1hr"
elif freq == "minutes":
    self.freq_suffix = "1min"
```

3. This step is straightforward as we simply load the market data into memory from the input source:

```
self.ticker_file_stream = os.path.join(
    f"{data_dir}",
    f"{'_'.join([self.exchange, self.ticker, \
                self.freq_suffix])}.csv",
)
assert os.path.isfile(
    self.ticker_file_stream
), f"Cryptocurrency exchange data file stream \
not found at: data/{self.ticker_file_stream}.csv"
# Cryptocurrency exchange data stream. An offline
# file stream is used. Alternatively, a web
# API can be used to pull live data.
self.ohlcv_df = pd.read_csv(
    self.ticker_file_stream,
    skiprows=1).sort_values(by="Date"
)

self.opening_account_balance = \
    env_config["opening_account_balance"]
```

4. Now, let's define the continuous action space and the observation space of the environment:

```python
self.action_space = spaces.Box(
    low=np.array([-1]), high=np.array([1]), \
                dtype=np.float
)

self.observation_features = [
    "Open",
    "High",
    "Low",
    "Close",
    "Volume BTC",
    "Volume USD",
]
self.obs_width, self.obs_height = 128, 128
self.horizon = env_config.get(
            "observation_horizon_sequence_length")
self.observation_space = spaces.Box(
    low=0, high=255, shape=(128, 128, 3),
    dtype=np.uint8,
)
```

5. Let's define the outline of the `step` method for the environment. We'll complete the helper method implementations in the following steps:

```python
def step(self, action):
    # Execute one step within the environment
    self.execute_trade_action(action)

    self.current_step += 1

    reward = self.account_value - \
        self.opening_account_balance  # Profit (loss)
    done = self.account_value <= 0 or \
            self.current_step >= len(
```

```
        self.ohlcv_df.loc[:, "Open"].values
    )


    obs = self.get_observation()


    return obs, reward, done, {}
```

6. The first helper method is the `execute_trade_action` method. The implementation in the next few steps should be straightforward, given that the previous recipes also implemented the logic behind buying and selling cryptocurrency at an exchange rate:

```
def execute_trade_action(self, action):

    if action == 0:  # Indicates "HODL" action
        # HODL position; No trade to be executed
        return
    order_type = "buy" if action > 0 else "sell"

    order_fraction_of_allowable_coins = abs(action)
    # Stochastically determine the current price
    # based on Market Open & Close
    current_price = random.uniform(
        self.ohlcv_df.loc[self.current_step, "Open"],
        self.ohlcv_df.loc[self.current_step,
                "Close"],
    )
```

7. A Buy order at the exchange can be simulated as follows:

```python
        if order_type == "buy":
            allowable_coins = \
                int(self.cash_balance / current_price)
            # Simulate a BUY order and execute it at
            # current_price
            num_coins_bought = int(allowable_coins * \
            order_fraction_of_allowable_coins)
            current_cost = self.cost_basis * \
                            self.num_coins_held
            additional_cost = num_coins_bought * \
                            current_price

            self.cash_balance -= additional_cost
            self.cost_basis = (current_cost + \
                                additional_cost) / (
                self.num_coins_held + num_coins_bought
            )
            self.num_coins_held += num_coins_bought

            if num_coins_bought > 0:
                self.trades.append(
                    {
                        "type": "buy",
                        "step": self.current_step,
                        "shares": num_coins_bought,
                        "proceeds": additional_cost,
                    }
                )
```

8. Similarly, a Sell order can be simulated in the following manner:

```python
elif order_type == "sell":
    # Simulate a SELL order and execute it at
    # current_price
    num_coins_sold = int(
        self.num_coins_held * \
        order_fraction_of_allowable_coins
    )
    self.cash_balance += num_coins_sold * \
                            current_price
    self.num_coins_held -= num_coins_sold
    sale_proceeds = num_coins_sold * \
                        current_price

    if num_coins_sold > 0:
        self.trades.append(
            {
                "type": "sell",
                "step": self.current_step,
                "shares": num_coins_sold,
                "proceeds": sale_proceeds,
            }
        )
```

9. Once the Buy/Sell order has been executed, the account balance needs to be updated:

```python
if self.num_coins_held == 0:
    self.cost_basis = 0
# Update account value
self.account_value = self.cash_balance + \
                        self.num_coins_held * \
                        current_price
```

10. To test `CryptoTradingVisualcontinuousEnv`, you can use the following
    lines of code for the __main__ function:

```
if __name__ == "__main__":
    env = CryptoTradingVisualContinuousEnv()
    obs = env.reset()
    for _ in range(600):
        action = env.action_space.sample()
        next_obs, reward, done, _ = env.step(action)
        env.render()
```

How it works...

`CryptoTradingVisualcontinuousEnv` provides an RL environment with a trader
screen-like image as the observation and provides a continuous, real-valued action space
for the Agents to act in. The actions in this environment are one-dimensional, continuous,
and real-valued and the magnitude indicates the fraction amount of the crypto coins/
tokens. If the action has a positive sign (0 to 1), it's interpreted as a Buy order, while if the
action has a negative sign (-1 to 0), it's interpreted as a Sell order. The fraction amount
is converted into a number of allowable coins that can be bought or sold based on the
balance in the trading account.

Training a cryptocurrency trading bot using RL

The soft actor-critic Agent is one of the most popular and state-of-the-art RL Agents
available and is based on an off-policy, maximum entropy-based deep RL algorithm. This
recipe provides all the ingredients you will need to build a soft actor-critic Agent from
scratch using TensorFlow 2.x and train it for cryptocurrency (Bitcoin, Ethereum, and so
on) trading using real data from the Gemini cryptocurrency exchange.

Getting ready

To complete this recipe, make sure you have the latest version. You will need to activate the `tf2rl-cookbook` Python/conda virtual environment. Make sure that you update the environment so that it matches the latest conda environment specification file (`tfrl-cookbook.yml`), which can be found in this cookbook's code repository. If the following `import` statements run without any issues, you are ready to get started:

```
mport functools
import os
import random
from collections import import deque
from functools import reduce

import imageio
import numpy as np
import tensorflow as tf
import tensorflow_probability as tfp
from tensorflow.keras.layers import Concatenate, Dense, Input
from tensorflow.keras.models import Model
from tensorflow.keras.optimizers import Adam

from crypto_trading_continuous_env import
CryptoTradingContinuousEnv
```

How to do it...

This recipe will guide you through the step-by-step process of implementing the SAC Agent. It will also help you train the agent in the cryptocurrency trading environments so that you can automate your profit-making machine!

Let's gear up and begin the implementation:

1. SAC is an actor-critic Agent, so it has both the actor and the critic components. Let's begin by defining our actor neural network using TensorFlow 2.x:

```
def actor(state_shape, action_shape, units=(512, 256,
64)):
    state_shape_flattened = \
        functools.reduce(lambda x, y: x * y, state_shape)
    state = Input(shape=state_shape_flattened)
```

```
    x = Dense(units[0], name="L0", activation="relu")\
            (state)
    for index in range(1, len(units)):
        x = Dense(units[index],name="L{}".format(index),\
                activation="relu")(x)

    actions_mean = Dense(action_shape[0], \
                    name="Out_mean")(x)
    actions_std = Dense(action_shape[0], \
                    name="Out_std")(x)

    model = Model(inputs=state,
                    outputs=[actions_mean, actions_std])

    return model
```

2. Next, let's define the critic neural network:

```
def critic(state_shape, action_shape, units=(512, 256,
64)):
    state_shape_flattened = \
        functools.reduce(lambda x, y: x * y, state_shape)
    inputs = [Input(shape=state_shape_flattened),
                Input(shape=action_shape)]
    concat = Concatenate(axis=-1)(inputs)
    x = Dense(units[0], name="Hidden0",
            activation="relu")(concat)
    for index in range(1, len(units)):
        x = Dense(units[index],
                name="Hidden{}".format(index),
                activation="relu")(x)

    output = Dense(1, name="Out_QVal")(x)
    model = Model(inputs=inputs, outputs=output)

    return model
```

3. Given the current model weights and the target model weights, let's implement a quick function that will slowly update the target weights using `tau` as the averaging factor. This is like the Polyak averaging step:

```python
def update_target_weights(model, target_model,
tau=0.005):
    weights = model.get_weights()
    target_weights = target_model.get_weights()
    for i in range(len(target_weights)):  # set tau% of
    # target model to be new weights
        target_weights[i] = weights[i] * tau + \
                        target_weights[i] * (1 - tau)
    target_model.set_weights(target_weights)
```

4. We are now ready to initialize our SAC Agent class:

```python
class SAC(object):
    def __init__(
        self,
        env,
        lr_actor=3e-5,
        lr_critic=3e-4,
        actor_units=(64, 64),
        critic_units=(64, 64),
        auto_alpha=True,
        alpha=0.2,
        tau=0.005,
        gamma=0.99,
        batch_size=128,
        memory_cap=100000,
    ):
        self.env = env
        self.state_shape = env.observation_space.shape
        # shape of observations
        self.action_shape = env.action_space.shape
        # number of actions
        self.action_bound = (env.action_space.high - \
```

```
                    env.action_space.low) / 2
self.action_shift = (env.action_space.high + \
                    env.action_space.low) / 2
self.memory = deque(maxlen=int(memory_cap))
```

5. As the next step, we'll initialize the actor network and print a summary of the actor neural network:

```
# Define and initialize actor network
self.actor = actor(self.state_shape,
                    self.action_shape, actor_units)
self.actor_optimizer = \
    Adam(learning_rate=lr_actor)
self.log_std_min = -20
self.log_std_max = 2
print(self.actor.summary())
```

6. Next, we'll define the two critic networks and print the summary of the critic neural network as well:

```
self.critic_1 = critic(self.state_shape,
                    self.action_shape, critic_units)
self.critic_target_1 = critic(self.state_shape,
                    self.action_shape, critic_units)
self.critic_optimizer_1 = \
        Adam(learning_rate=lr_critic)
update_target_weights(self.critic_1, \
                    self.critic_target_1, tau=1.0)

self.critic_2 = critic(self.state_shape, \
                    self.action_shape, critic_units)
self.critic_target_2 = critic(self.state_shape,\
                    self.action_shape, critic_units)
self.critic_optimizer_2 = \
        Adam(learning_rate=lr_critic)
update_target_weights(self.critic_2, \
```

```
        self.critic_target_2, tau=1.0)

    print(self.critic_1.summary())
```

7. Let's initialize the `alpha` temperature parameter and the target entropy:

```
    self.auto_alpha = auto_alpha
    if auto_alpha:
        self.target_entropy = \
            -np.prod(self.action_shape)
        self.log_alpha = \
            tf.Variable(0.0, dtype=tf.float64)
        self.alpha = \
            tf.Variable(0.0, dtype=tf.float64)
        self.alpha.assign(tf.exp(self.log_alpha))
        self.alpha_optimizer = \
            Adam(learning_rate=lr_actor)
    else:
        self.alpha = tf.Variable(alpha,
                                 dtype=tf.float64)
```

8. We'll also initialize the other hyperparameters of SAC:

```
    self.gamma = gamma  # discount factor
    self.tau = tau  # target model update
    self.batch_size = batch_size
```

9. That completes the __init__ method of the SAC agent. Next, we'll implement a method that will (pre)process the action that's taken:

```
def process_actions(self, mean, log_std, test=False,
eps=1e-6):
    std = tf.math.exp(log_std)
    raw_actions = mean

    if not test:
        raw_actions += tf.random.normal(shape=mean.\
                        shape, dtype=tf.float64) * std

    log_prob_u = tfp.distributions.Normal(loc=mean,
                    scale=std).log_prob(raw_actions)
    actions = tf.math.tanh(raw_actions)

    log_prob = tf.reduce_sum(log_prob_u - \
                tf.math.log(1 - actions ** 2 + eps))

    actions = actions * self.action_bound + \
                self.action_shift

    return actions, log_prob
```

10. We are now ready to implement the act method in order to generate the SAC agent's action, given a state:

```
def act(self, state, test=False, use_random=False):
    state = state.reshape(-1)  # Flatten state
    state = \
    np.expand_dims(state, axis=0).astype(np.float64)

    if use_random:
        a = tf.random.uniform(
                shape=(1, self.action_shape[0]), \
                minval=-1, maxval=1, dtype=tf.float64
        )
    else:
```

```
        means, log_stds = self.actor.predict(state)
        log_stds = tf.clip_by_value(log_stds,
                                            self.log_std_min,
                                            self.log_std_max)


        a, log_prob = self.process_actions(means,
                                                log_stds,
                                                test=test)


        q1 = self.critic_1.predict([state, a])[0][0]
        q2 = self.critic_2.predict([state, a])[0][0]
        self.summaries["q_min"] = tf.math.minimum(q1, q2)
        self.summaries["q_mean"] = np.mean([q1, q2])


        return a
```

11. In order to save experiences to the Replay memory, let's implement the `remember` function:

```
def remember(self, state, action, reward, next_state,
done):
        state = state.reshape(-1)  # Flatten state
        state = np.expand_dims(state, axis=0)
        next_state = next_state.reshape(-1)
        # Flatten next-state
        next_state = np.expand_dims(next_state, axis=0)
        self.memory.append([state, action, reward,
                            next_state, done])
```

12. Now, let's begin implementing the experience replay process. We'll start by initializing the replay method. We'll complete the implementation of the replay method in the upcoming steps:

```
def replay(self):
        if len(self.memory) < self.batch_size:
            return


        samples = random.sample(self.memory, self.batch_
```

```
   size)
           s = np.array(samples).T
           states, actions, rewards, next_states, dones = [
               np.vstack(s[i, :]).astype(np.float) for i in\
               range(5)
           ]
```

13. Let's start a persistent `GradientTape` function and begin accumulating gradients. We'll do this by processing the actions and obtaining the next set of actions and log probabilities:

```
with tf.GradientTape(persistent=True) as tape:
    # next state action log probs
    means, log_stds = self.actor(next_states)
    log_stds = tf.clip_by_value(log_stds,
                                self.log_std_min,
                                self.log_std_max)
    next_actions, log_probs = \
        self.process_actions(means, log_stds)
```

14. With that, we can now compute the losses of the two critic networks:

```
    current_q_1 = self.critic_1([states,
                                 actions])
    current_q_2 = self.critic_2([states,
                                 actions])
    next_q_1 = self.critic_target_1([next_states,
                                     next_actions])
    next_q_2 = self.critic_target_2([next_states,
                                     next_actions])
    next_q_min = tf.math.minimum(next_q_1,
                                 next_q_2)
    state_values = next_q_min - self.alpha * \
                                log_probs
    target_qs = tf.stop_gradient(
        rewards + state_values * self.gamma * \
        (1.0 - dones)
```

```
    )
    critic_loss_1 = tf.reduce_mean(
        0.5 * tf.math.square(current_q_1 - \
                            target_qs)
    )
    critic_loss_2 = tf.reduce_mean(
        0.5 * tf.math.square(current_q_2 - \
                            target_qs)
    )
```

15. The current state-action and log probabilities, as prescribed by the actor, can be computed as follows:

```
    means, log_stds = self.actor(states)
    log_stds = tf.clip_by_value(log_stds,
                                self.log_std_min,
                                self.log_std_max)
    actions, log_probs = \
        self.process_actions(means, log_stds)
```

16. We can now compute the actor loss and apply gradients to the critic:

```
    current_q_1 = self.critic_1([states,
                                actions])
    current_q_2 = self.critic_2([states,
                                actions])
    current_q_min = tf.math.minimum(current_q_1,
                                    current_q_2)
    actor_loss = tf.reduce_mean(self.alpha * \
                    log_probs - current_q_min)
    if self.auto_alpha:
        alpha_loss = -tf.reduce_mean(
            (self.log_alpha * \
            tf.stop_gradient(log_probs + \
                            self.target_entropy))
        )
    critic_grad = tape.gradient(
        critic_loss_1,
```

```
        self.critic_1.trainable_variables
    )
    self.critic_optimizer_1.apply_gradients(
        zip(critic_grad,
        self.critic_1.trainable_variables)
    )
```

17. Similarly, we can compute and apply the actor's gradients:

```
    critic_grad = tape.gradient(
        critic_loss_2,
    self.critic_2.trainable_variables
    )  # compute actor gradient
    self.critic_optimizer_2.apply_gradients(
        zip(critic_grad,
        self.critic_2.trainable_variables)
    )

    actor_grad = tape.gradient(
        actor_loss, self.actor.trainable_variables
    )  # compute actor gradient
    self.actor_optimizer.apply_gradients(
        zip(actor_grad,
            self.actor.trainable_variables)
    )
```

18. Now, let's log the summaries to TensorBoard:

```
    # tensorboard info
    self.summaries["q1_loss"] = critic_loss_1
    self.summaries["q2_loss"] = critic_loss_2
    self.summaries["actor_loss"] = actor_loss

    if self.auto_alpha:
        # optimize temperature
        alpha_grad = tape.gradient(alpha_loss,
                                   [self.log_alpha])
```

```
self.alpha_optimizer.apply_gradients(
                zip(alpha_grad, [self.log_alpha]))
self.alpha.assign(tf.exp(self.log_alpha))
# tensorboard info
self.summaries["alpha_loss"] = alpha_loss
```

19. That completes our experience replay method. Now, we can move on to the `train` method's implementation. Let's begin by initializing the `train` method. We will complete the implementation of this method in the following steps:

```
def train(self, max_epochs=8000, random_epochs=1000,
max_steps=1000, save_freq=50):
    current_time = datetime.datetime.now().\
                    strftime("%Y%m%d-%H%M%S")
    train_log_dir = os.path.join("logs",
            "TFRL-Cookbook-Ch4-SAC", current_time)
    summary_writer = \
        tf.summary.create_file_writer(train_log_dir)

    done, use_random, episode, steps, epoch, \
    episode_reward = (
        False,
        True,
        0,
        0,
        0,
        0,
    )
    cur_state = self.env.reset()
```

20. Now, we are ready to start the main training loop. First, let's handle the end of episode case:

```
while epoch < max_epochs:
    if steps > max_steps:
        done = True
```

```python
        if done:
            episode += 1
            print(
                "episode {}: {} total reward,
                {} alpha, {} steps,
                {} epochs".format(
                    episode, episode_reward,
                    self.alpha.numpy(), steps, epoch
                )
            )

            with summary_writer.as_default():
                tf.summary.scalar(
                    "Main/episode_reward", \
                        episode_reward, step=episode
                )
                tf.summary.scalar(
                    "Main/episode_steps",
                    steps, step=episode)
            summary_writer.flush()

            done, cur_state, steps, episode_reward =\
                False, self.env.reset(), 0, 0
            if episode % save_freq == 0:
                self.save_model(
                    "sac_actor_episode{}.h5".\
                        format(episode),
                    "sac_critic_episode{}.h5".\
                        format(episode),
                )
```

21. For every step into the environment, the following steps will need to be executed for the SAC agent to learn:

```python
            if epoch > random_epochs and \
                len(self.memory) > self.batch_size:
```

```
            use_random = False

    action = self.act(cur_state, \
        use_random=use_random)  # determine action
    next_state, reward, done, _ = \
        self.env.step(action[0])  # act on env
    # self.env.render(mode='rgb_array')

    self.remember(cur_state, action, reward,
                next_state, done)  #add to memory
    self.replay()  # train models through memory
    # replay

    update_target_weights(
        self.critic_1, self.critic_target_1,
        tau=self.tau
    )  # iterates target model
    update_target_weights(self.critic_2,
    self.critic_target_2,
    tau=self.tau)

    cur_state = next_state
    episode_reward += reward
    steps += 1
    epoch += 1
```

22. With the agent updates taken care of, we can now log some more useful information to TensorBoard:

```
    # Tensorboard update
    with summary_writer.as_default():
        if len(self.memory) > self.batch_size:
            tf.summary.scalar(
                "Loss/actor_loss",
                self.summaries["actor_loss"],
                step=epoch
            )
```

```
            tf.summary.scalar(
                "Loss/q1_loss",
                 self.summaries["q1_loss"],
                 step=epoch
            )
            tf.summary.scalar(
                "Loss/q2_loss",
                 self.summaries["q2_loss"],
                 step=epoch
            )
            if self.auto_alpha:
                tf.summary.scalar(
                    "Loss/alpha_loss",
                     self.summaries["alpha_loss"],
                     step=epoch
                )

        tf.summary.scalar("Stats/alpha",
                           self.alpha, step=epoch)
        if self.auto_alpha:
            tf.summary.scalar("Stats/log_alpha",
                          self.log_alpha, step=epoch)
        tf.summary.scalar("Stats/q_min",
            self.summaries["q_min"], step=epoch)
        tf.summary.scalar("Stats/q_mean",
            self.summaries["q_mean"], step=epoch)
        tf.summary.scalar("Main/step_reward",
                           reward, step=epoch)
    summary_writer.flush()
```

23. As the last step in our train method implementation, we can save the actor and critic models to facilitate resuming our training or reloading from a checkpoint:

```
self.save_model(
        "sac_actor_final_episode{}.h5".
format(episode),
        "sac_critic_final_episode{}.h5".
format(episode),
    )
```

24. Now, we'll actually implement the save_model method we referenced previously:

```
def save_model(self, a_fn, c_fn):
    self.actor.save(a_fn)
    self.critic_1.save(c_fn)
```

25. Let's quickly implement a method that will load the actor and critic states from the saved model so that we can restore/resume from a previously saved checkpoint when needed:

```
def load_actor(self, a_fn):
    self.actor.load_weights(a_fn)
    print(self.actor.summary())


def load_critic(self, c_fn):
    self.critic_1.load_weights(c_fn)
    self.critic_target_1.load_weights(c_fn)
    self.critic_2.load_weights(c_fn)
    self.critic_target_2.load_weights(c_fn)
    print(self.critic_1.summary())
```

26. To run the SAC agent in "test" mode, we can implement a helper method:

```
def test(self, render=True, fps=30,
filename="test_render.mp4"):
    cur_state, done, rewards = self.env.reset(), \
                               False, 0
    video = imageio.get_writer(filename, fps=fps)
```

```
    while not done:
        action = self.act(cur_state, test=True)
        next_state, reward, done, _ = \
                            self.env.step(action[0])
        cur_state = next_state
        rewards += reward
        if render:
            video.append_data(
                self.env.render(mode="rgb_array"))
    video.close()
    return rewards
```

27. That completes our SAC agent implementation. We are now ready to train the SAC agent in `CryptoTradingContinuousEnv`:

```
if __name__ == "__main__":
    gym_env = CryptoTradingContinuousEnv()
    sac = SAC(gym_env)
    # Load Actor and Critic from previously saved
    # checkpoints
    # sac.load_actor("sac_actor_episodexyz.h5")
    # sac.load_critic("sac_critic_episodexyz.h5")
    sac.train(max_epochs=100000, random_epochs=10000,
            save_freq=50)
    reward = sac.test()
    print(reward)
```

How it works...

SAC is a powerful RL algorithm and has proven to be effective across a variety of RL simulation environments. SAC maximizes the entropy of the agent's policy, in addition to optimizing for the maximum episodic rewards. You can watch the progress of the agent as it learns to trade using the TensorBoard since this recipe includes code for logging the agent's progress along the way. You can launch TensorBoard using the following command:

```
tensorboard --logdir=logs
```

The preceding command will launch TensorBoard. You can access it with your browser at the default address of http://localhost:6006. A sample TensorBoard screenshot has been provided here for reference:

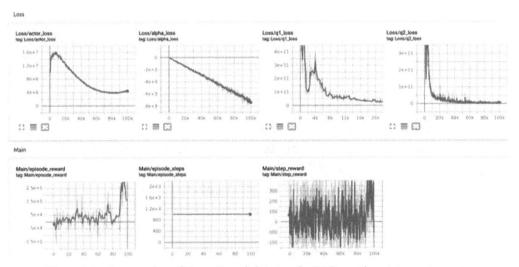

Figure 4.4 – A screenshot of TensorBoard showing the SAC agent's training progress in CryptoTradingContinuousEnv

That concludes this recipe and this chapter. Happy training!

5

Reinforcement Learning in the Real World – Building Stock/Share Trading Agents

Software-based **deep reinforcement learning (deep RL)** agents have tremendous potential when it comes to executing trading strategies tirelessly and flawlessly without limitations based on memory capacity, speed, efficiency, and emotional disturbances that a human trader is prone to facing. Profitable trading in the stock market involves carefully executing buy/sell trades with stock symbols/tickers while taking into account several market factors such as trading conditions and macro and micro market conditions, in addition to social, political, and company-specific changes. Deep RL agents have a lot of potential when it comes to solving challenging problems in the real world and a lot of opportunities exist.

However, only a few successful stories of using deep RL agents in the real world beyond games exist due to the various challenges associated with real-world deployments of RL agents. This chapter contains recipes that will help you successfully develop RL agents for yet another interesting and rewarding real-world problem: stock market trading. The recipes provided contain information on how to implement custom OpenAI Gym-compatible learning environments for stock market trading with both discrete and continuous value action spaces. In addition, you will learn how to build and train RL agents for stock trading learning environments.

Specifically, the following recipes will be covered in this chapter:

- Building a stock market trading RL platform using real stock exchange data

- Building a stock market trading RL platform using price charts

- Building an advanced stock trading RL platform to train agents to mimic professional traders

Let's get started!

Technical requirements

The code in this book has been extensively tested on Ubuntu 18.04 and Ubuntu 20.04 and should work with later versions of Ubuntu if Python 3.6+ is available. With Python 3.6+ installed, along with the necessary Python packages, as listed at the start of each of the recipes, the code should run fine on Windows and Mac OS X too. You should create and use a Python virtual environment named `tf2rl-cookbook` to install the packages and run the code in this book. Installing Miniconda or Anaconda for Python virtual environment management is recommended.

The complete code for each recipe in each chapter is available here: `https://github.com/PacktPublishing/Tensorflow-2-Reinforcement-Learning-Cookbook`.

Building a stock market trading RL platform using real stock exchange data

The stock market provides anyone with a highly lucrative opportunity to participate and make profits. While it is easily accessible, not all humans can make consistently profitable trades due to the dynamic nature of the market and the emotional aspects that can impair people's actions. RL agents take emotion out of the equation and can be trained to make profits consistently. This recipe will teach you how to implement a stock market trading environment that will teach your RL agents how to trade stocks using real stock market data. When you have trained them enough, you can deploy them so that they automatically make trades (and profits) for you!

Getting ready

To complete this recipe, make sure you have the latest version. You will need to activate the tf2rl-cookbook Python/conda virtual environment. Make sure you update the environment so that it matches the latest conda environment specification file (tfrl-cookbook.yml), which can be found in this cookbook's code repository. If the following import statements run without any issues, you are ready to get started:

```
import os
import random
from typing import Dict

import gym
import numpy as np
import pandas as pd
from gym import spaces

from trading_utils import TradeVisualizer
```

How to do it...

Follow this step-by-step process to implement `StockTradingEnv`:

1. Let's initialize the configurable parameters of the environment:

    ```python
    env_config = {
        "ticker": "TSLA",
        "opening_account_balance": 1000,
        # Number of steps (days) of data provided to the
        # agent in one observation
        "observation_horizon_sequence_length": 30,
        "order_size": 1,   # Number of shares to buy per
        # buy/sell order
    }
    ```

2. Let's initialize the `StockTradingEnv` class and load the stock market data for the configured stock ticker symbol:

    ```python
    class StockTradingEnv(gym.Env):
        def __init__(self, env_config: Dict = env_config):
            """Stock trading environment for RL agents
            Args:
                ticker (str, optional): Ticker symbol for the
                stock. Defaults to "MSFT".
                env_config (Dict): Env configuration values
            """
            super(StockTradingEnv, self).__init__()
            self.ticker = env_config.get("ticker", "MSFT")
            data_dir = os.path.join(os.path.dirname(os.path.\
                            realpath(__file__)), "data")
            self.ticker_file_stream = os.path.join(f"{
                            data_dir}", f"{self.ticker}.csv")
    ```

3. Let's make sure that the stock market data source exists and then load the data stream:

    ```python
            assert os.path.isfile(
                self.ticker_file_stream
    ```

```
    ), f"Historical stock data file stream not found
     at: data/{self.ticker}.csv"
    # Stock market data stream. An offline file
    # stream is used. Alternatively, a web
    # API can be used to pull live data.
    # Data-Frame: Date Open High Low Close Adj-Close
    # Volume
    self.ohlcv_df = \
        pd.read_csv(self.ticker_file_stream)
```

4. We are now ready to define the observation and action space/environment in order to complete our initialization function definition:

```
    self.opening_account_balance = \
        env_config["opening_account_balance"]
    # Action: 0-> Hold; 1-> Buy; 2 ->Sell;
    self.action_space = spaces.Discrete(3)

    self.observation_features = [
        "Open",
        "High",
        "Low",
        "Close",
        "Adj Close",
        "Volume",
    ]
    self.horizon = env_config.get(
            "observation_horizon_sequence_length")
    self.observation_space = spaces.Box(
        low=0,
        high=1,
        shape=(len(self.observation_features),
            self.horizon + 1),
        dtype=np.float,
    )
    self.order_size = env_config.get("order_size")
```

5. Next, we will implement a method so that we can gather observations:

```python
def get_observation(self):
    # Get stock price info data table from input
    # (file/live) stream
    observation = (
        self.ohlcv_df.loc[
            self.current_step : self.current_step + \
                self.horizon,
            self.observation_features,
        ]
        .to_numpy()
        .T
    )
    return observation
```

6. Next, to execute a trade order, we need the required logic to be in place. Let's add this now:

```python
def execute_trade_action(self, action):
    if action == 0:  # Hold position
        return
    order_type = "buy" if action == 1 else "sell"

    # Stochastically determine the current stock
    # price based on Market Open & Close
    current_price = random.uniform(
        self.ohlcv_df.loc[self.current_step, "Open"],
        self.ohlcv_df.loc[self.current_step,
                        "Close"],
    )
```

7. With the initialization done, we can add the logic for buying stock:

```python
if order_type == "buy":
    allowable_shares = \
        int(self.cash_balance / current_price)
```

```python
        if allowable_shares < self.order_size:
            # Not enough cash to execute a buy order
            # return
        # Simulate a BUY order and execute it at
        # current_price
        num_shares_bought = self.order_size
        current_cost = self.cost_basis * \
                        self.num_shares_held
        additional_cost = num_shares_bought * \
                        current_price

        self.cash_balance -= additional_cost
        self.cost_basis = (current_cost + \
                            additional_cost) / (
            self.num_shares_held + num_shares_bought
        )
        self.num_shares_held += num_shares_bought

        self.trades.append(
            {
                "type": "buy",
                "step": self.current_step,
                "shares": num_shares_bought,
                "proceeds": additional_cost,
            }
        )
```

8. Likewise, we can now add the logic for selling stock:

```python
    elif order_type == "sell":
        # Simulate a SELL order and execute it at
        # current_price
        if self.num_shares_held < self.order_size:
            # Not enough shares to execute a sell
            # order
            return
        num_shares_sold = self.order_size
```

```
            self.cash_balance += num_shares_sold * \
                                    current_price
            self.num_shares_held -= num_shares_sold
            sale_proceeds = num_shares_sold * current_
price

            self.trades.append(
                {
                    "type": "sell",
                    "step": self.current_step,
                    "shares": num_shares_sold,
                    "proceeds": sale_proceeds,
                }
            )
```

9. Finally, let's update the account balance:

```
    # Update account value
    self.account_value = self.cash_balance + \
                            self.num_shares_held * \
                            current_price
```

10. We are now ready to fire and check out the new environment:

```
if __name__ == "__main__":
    env = StockTradingEnv()
    obs = env.reset()
    for _ in range(600):
        action = env.action_space.sample()
        next_obs, reward, done, _ = env.step(action)
        env.render()
```

How it works...

The observations are stock price information (OHLCV) over a horizon, as specified in env_config. The action space is discrete to allow us to perform buy/sell/hold trades. This is a starter environment for RL agents to learn to trade stocks in the stock market. Happy training!

Building a stock market trading RL platform using price charts

Human traders look at several indicators on their price monitor in order to vet and identify a potential trade. Can we allow the agents to also visually look at the price candlestick charts to trade stocks instead of providing just a tabular/CSV representation? Yes, we can! This recipe will teach you how to build a visually rich trading environment for your RL agents.

Getting ready

To complete this recipe, make sure you have the latest version. You will need to activate the tf2rl-cookbook Python/conda virtual environment. Make sure that you update the environment so that it matches the latest conda environment specification file (tfrl-cookbook.yml), which can be found in this cookbook's code repository. If the following import statements run without any issues, you are ready to get started:

```
import os
import random
from typing import Dict

import cv2
import gym
import numpy as np
import pandas as pd
from gym import spaces

from trading_utils import TradeVisualizer
```

How to do it...

Let's start by configuring the environment. Then, we will guide you through the process of completing the implementation. By the end of this recipe, you will have built a complete stock trading RL environment that allows an agent to process visual stock charts and make trading decisions.

Let's get started:

1. Configure the learning environment, as follows:

```python
env_config = {
    "ticker": "TSLA",
    "opening_account_balance": 100000,
    # Number of steps (days) of data provided to the
    # agent in one observation
    "observation_horizon_sequence_length": 30,
    "order_size": 1,   # Number of shares to buy per
    # buy/sell order
}
```

2. Let's implement the initialization step for `StockTradingVisualEnv`:

```python
class StockTradingVisualEnv(gym.Env):
    def __init__(self, env_config: Dict = env_config):
        """Stock trading environment for RL agents

        Args:
            ticker (str, optional): Ticker symbol for the
            stock. Defaults to "MSFT".
            env_config (Dict): Env configuration values
        """
        super(StockTradingVisualEnv, self).__init__()
        self.ticker = env_config.get("ticker", "MSFT")
        data_dir = os.path.join(os.path.dirname(os.path.\
                        realpath(__file__)), "data")
        self.ticker_file_stream = os.path.join(
                f"{data_dir}", f"{self.ticker}.csv")
        assert os.path.isfile(
            self.ticker_file_stream
        ), f"Historical stock data file stream not found\
            at: data/{self.ticker}.csv"
        # Stock market data stream. An offline file
        # stream is used. Alternatively, a web
        # API can be used to pull live data.
```

```
# Data-Frame: Date Open High Low Close Adj-Close
# Volume
self.ohlcv_df = \
    pd.read_csv(self.ticker_file_stream)
```

3. Let's complete the implementation of the __init__ method:

```
self.opening_account_balance = \
    env_config["opening_account_balance"]

self.action_space = spaces.Discrete(3)

self.observation_features = [
    "Open",
    "High",
    "Low",
    "Close",
    "Adj Close",
    "Volume",
]
self.obs_width, self.obs_height = 128, 128
self.horizon = env_config.get(
    "observation_horizon_sequence_length")
self.observation_space = spaces.Box(
    low=0, high=255, shape=(128, 128, 3),
    dtype=np.uint8,
)
self.order_size = env_config.get("order_size")
self.viz = None  # Visualizer
```

4. The next step to is the define the step method for the environment:

```
def step(self, action):
    # Execute one step within the trading environment
    self.execute_trade_action(action)
    self.current_step += 1
    reward = self.account_value - \
```

```
                self.opening_account_balance  # Profit (loss)
        done = self.account_value <= 0 or \
                self.current_step >= len(
            self.ohlcv_df.loc[:, "Open"].values
        )
        obs = self.get_observation()
        return obs, reward, done, {}
```

5. Let's implement the two missing methods we used in the preceding
    step. To implement the get_observation method, we will need the
    TradeVisualizer method to be initialized. Due to this, let's implement the
    reset method first:

```
def reset(self):
    # Reset the state of the environment to an
    # initial state
    self.cash_balance = self.opening_account_balance
    self.account_value = self.opening_account_balance
    self.num_shares_held = 0
    self.cost_basis = 0
    self.current_step = 0
    self.trades = []
    if self.viz is None:
        self.viz = TradeVisualizer(
            self.ticker,
            self.ticker_file_stream,
            "TFRL-Cookbook Ch4-StockTradingVisualEnv",
        )

    return self.get_observation()
```

6. Now, let's continue with our implementation of the `get_observation` method:

```
def get_observation(self):
    """Return a view of the Ticker price chart as
        image observation
    Returns:
        img_observation (np.ndarray): Image of ticker
        candle stick plot with volume bars as
        observation
    """
    img_observation = \
        self.viz.render_image_observation(
            self.current_step, self.horizon
        )
    img_observation = cv2.resize(
        img_observation, dsize=(128, 128),
        interpolation=cv2.INTER_CUBIC
    )
    return img_observation
```

7. It's time to implement the logic that will execute the trade actions taken by the agent. We'll split our implementation of the trade execution logic into the next three steps:

```
def execute_trade_action(self, action):
    if action == 0:  # Hold position
        return
    order_type = "buy" if action == 1 else "sell"

    # Stochastically determine the current stock
    # price based on Market Open & Close
    current_price = random.uniform(
        self.ohlcv_df.loc[self.current_step, "Open"],
        self.ohlcv_df.loc[self.current_step, \
                    "Close"],
    )
```

8. Let's implement the logic to execute "buy" orders:

```
if order_type == "buy":
        allowable_shares = \
           int(self.cash_balance / current_price)
        if allowable_shares < self.order_size:
           return
        num_shares_bought = self.order_size
        current_cost = self.cost_basis * \
                       self.num_shares_held
        additional_cost = num_shares_bought * \
                       current_price
        self.cash_balance -= additional_cost
        self.cost_basis = (current_cost + \
                       additional_cost)/ \
                       (self.num_shares_held +\
                       num_shares_bought)
        self.num_shares_held += num_shares_bought
        self.trades.append(
           {   "type": "buy",
               "step": self.current_step,
               "shares": num_shares_bought,
               "proceeds": additional_cost,
           }
        )
```

9. Now, let's take care of handling "sell" orders:

```
elif order_type == "sell":
        # Simulate a SELL order and execute it at
        # current_price
        if self.num_shares_held < self.order_size:
           # Not enough shares to execute a sell
           # order
           return
        num_shares_sold = self.order_size
        self.cash_balance += num_shares_sold * \
                       current_price
```

```
            self.num_shares_held -= num_shares_sold
            sale_proceeds = num_shares_sold * \
                            current_price
            self.trades.append(
                {
                    "type": "sell",
                    "step": self.current_step,
                    "shares": num_shares_sold,
                    "proceeds": sale_proceeds,
                }
            )
        if self.num_shares_held == 0:
            self.cost_basis = 0
        # Update account value
        self.account_value = self.cash_balance + \
                             self.num_shares_held * \
                             current_price
```

10. With that, our implementation is complete! We are now ready to test the environment using a randomly acting agent:

```
if __name__ == "__main__":
    env = StockTradingVisualEnv()
    obs = env.reset()
    for _ in range(600):
        action = env.action_space.sample()
        next_obs, reward, done, _ = env.step(action)
        env.render()
```

How it works...

The observations in `StockTradingVisualEnv` are stock price information (OHLCV) over a horizon, as specified in `env_config`. The action space is discrete so that we can buy/sell/hold trades. More specifically, the action has the following meaning: 0-> Hold; 1-> Buy; 2 ->Sell.

The following image illustrates the environment in action:

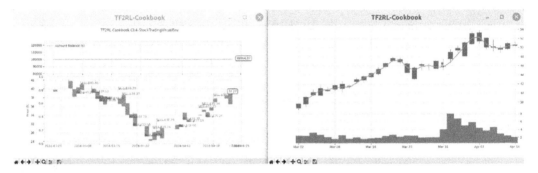

Figure 5.1 – A sample screenshot of StockTradingVisualEnv in action

Building an advanced stock trading RL platform to train agents to mimic professional traders

This recipe will help you implement a complete stock trading environment with high-dimensional image observation spaces and continuous action spaces for training your RL and deep RL agents. This will allow you to build intelligent trading bots using RL so that you can closely approximate how a professional stock trader would trade stocks. Like a professional trader, the RL agent you'll train will watch the stock market data in the form of candlesticks and price line charts and make trading decisions. A well-trained RL agent could potentially make thousands of profitable trades without needing a break or commission, unlike a human professional trader, thereby increasing your profits!

Getting ready

To complete this recipe, make sure you have the latest version. You will need to activate the `tf2rl-cookbook` Python/conda virtual environment. Make sure that you update the environment so that it matches the latest conda environment specification file (`tfrl-cookbook.yml`), which can be found in this cookbook's code repository. If the following `import` statements run without any issues, you are ready to get started:

```
import os
import random
from typing import Dict

import cv2
import gym
import numpy as np
import pandas as pd
from gym import spaces

from trading_utils import TradeVisualizer
```

How to do it...

By now, you must be familiar with the basic flow of the implementation from having worked through the previous recipes in this chapter. Follow these steps to put together a complete stock trading environment from scratch for training your advanced RL agents:

1. Let's begin our implementation of `StockTradingVisualContinuousEnv`:

    ```
    def __init__(self, env_config: Dict = env_config):
        """Stock trading environment for RL agents with
            continuous action space

        Args:
            ticker (str, optional): Ticker symbol for the
            stock. Defaults to "MSFT".
            env_config (Dict): Env configuration values
        """
        super(StockTradingVisualContinuousEnv,
            self).__init__()
        self.ticker = env_config.get("ticker", "MSFT")
    ```

```python
        data_dir = os.path.join(os.path.dirname(os.path.\
                        realpath(__file__)), "data")
        self.ticker_file_stream = os.path.join(
                    f"{data_dir}", f"{self.ticker}.csv")
        assert os.path.isfile(
            self.ticker_file_stream
        ), f"Historical stock data file stream not found
            at: data/{self.ticker}.csv"

        self.ohlcv_df = \
            pd.read_csv(self.ticker_file_stream)
```

2. Let's define the state space, action space, and other essential variables to complete the __init__ function's implementation:

```python
        self.opening_account_balance = \
            env_config["opening_account_balance"]
        # Action: 1-dim value indicating a fraction
        # amount of shares to Buy (0 to 1) or
        # sell (-1 to 0). The fraction is taken on the
        # allowable number of
        # shares that can be bought or sold based on the
        # account balance (no margin).
        self.action_space = spaces.Box(
            low=np.array([-1]), high=np.array([1]),
                dtype=np.float
        )

        self.observation_features = [
            "Open",
            "High",
            "Low",
            "Close",
            "Adj Close",
            "Volume",
        ]
        self.obs_width, self.obs_height = 128, 128
```

```
self.horizon = env_config.get(
    "observation_horizon_sequence_length")
self.observation_space = spaces.Box(
    low=0, high=255, shape=(128, 128, 3),
    dtype=np.uint8,
)
self.viz = None  # Visualizer
```

3. Next, let's implement the get_observation method:

```
def get_observation(self):
    """Return a view of the Ticker price chart as
        image observation

    Returns:
        img_observation (np.ndarray): Image of ticker
        candle stick plot with volume bars as
        observation
    """
    img_observation = \
        self.viz.render_image_observation(
        self.current_step, self.horizon
    )
    img_observation = cv2.resize(
        img_observation, dsize=(128, 128),
        interpolation=cv2.INTER_CUBIC
    )

    return img_observation
```

4. Let's initialize the trade execution logic:

```
def execute_trade_action(self, action):

    if action == 0:  # Indicates "Hold" action
        # Hold position; No trade to be executed
```

```
        return

    order_type = "buy" if action > 0 else "sell"

    order_fraction_of_allowable_shares = abs(action)
    # Stochastically determine the current stock
    # price based on Market Open & Close
    current_price = random.uniform(
        self.ohlcv_df.loc[self.current_step, "Open"],
        self.ohlcv_df.loc[self.current_step,
                        "Close"],
    )
```

5. We are now ready to define the behavior of a "buy" action:

```
    if order_type == "buy":
        allowable_shares = \
            int(self.cash_balance / current_price)
        # Simulate a BUY order and execute it at
        # current_price
        num_shares_bought = int(
            allowable_shares * \
                order_fraction_of_allowable_shares
        )
        current_cost = self.cost_basis * \
                        self.num_shares_held
        additional_cost = num_shares_bought * \
                        current_price

        self.cash_balance -= additional_cost
        self.cost_basis = (current_cost + \
                        additional_cost) / (
            self.num_shares_held + num_shares_bought
        )
        self.num_shares_held += num_shares_bought
```

```
if num_shares_bought > 0:
    self.trades.append(
        {
            "type": "buy",
            "step": self.current_step,
            "shares": num_shares_bought,
            "proceeds": additional_cost,
        }
    )
```

6. Similarly, we can define the behavior of the `"sell"` action and update the account balance to finalize the method's implementation:

```
elif order_type == "sell":
    # Simulate a SELL order and execute it at
    # current_price
    num_shares_sold = int(
        self.num_shares_held * \
        order_fraction_of_allowable_shares
    )
    self.cash_balance += num_shares_sold * \
                        current_price
    self.num_shares_held -= num_shares_sold
    sale_proceeds = num_shares_sold * \
                    current_price

    if num_shares_sold > 0:
        self.trades.append(
            {
                "type": "sell",
                "step": self.current_step,
                "shares": num_shares_sold,
                "proceeds": sale_proceeds,
            }
        )
    if self.num_shares_held == 0:
```

```
        self.cost_basis = 0
    # Update account value
    self.account_value = self.cash_balance + \
                        self.num_shares_held * \
                        current_price
```

7. We are now ready to implement the `step` method, which allows the agent to step through the environment:

```
def step(self, action):
    # Execute one step within the environment
    self.execute_trade_action(action)

    self.current_step += 1

    reward = self.account_value - \
        self.opening_account_balance  # Profit (loss)
    done = self.account_value <= 0 or \
        self.current_step >= len(
        self.ohlcv_df.loc[:, "Open"].values
    )

    obs = self.get_observation()

    return obs, reward, done, {}
```

8. Next, let's implement the `reset()` method, which will be executed at the start of every episode:

```
def reset(self):
    # Reset the state of the environment to an
    # initial state
    self.cash_balance = self.opening_account_balance
    self.account_value = self.opening_account_balance
    self.num_shares_held = 0
    self.cost_basis = 0
    self.current_step = 0
    self.trades = []
```

```python
        if self.viz is None:
            self.viz = TradeVisualizer(
                self.ticker,
                self.ticker_file_stream,
                "TFRL-Cookbook \
                Ch4-StockTradingVisualContinuousEnv",
            )

        return self.get_observation()
```

9. Let's finalize our implementation of the environment by implementing the `render` and `close` methods:

```python
    def render(self, **kwargs):
        # Render the environment to the screen

        if self.current_step > self.horizon:
            self.viz.render(
                self.current_step,
                self.account_value,
                self.trades,
                window_size=self.horizon,
            )

    def close(self):
        if self.viz is not None:
            self.viz.close()
            self.viz = None
```

10. Now, it's time for you to get one of the agents you built as part of the previous chapter to train and test this real data-backed stock market trading environment. For now, let's test the environment with a simple, random agent:

```python
if __name__ == "__main__":
    env = StockTradingVisualContinuousEnv()
    obs = env.reset()
    for _ in range(600):
```

```
action = env.action_space.sample()
next_obs, reward, done, _ = env.step(action)
env.render()
```

How it works...

To simulate a stock market, a real stock market data stream must be used. An offline file-based stream is utilized as an alternative to a web-based API, which would require internet connectivity and potentially a user account to fetch market data. The file stream contains the market data in a standard format: Date, Open, High, Low, Close, Adj-Close, and Volume.

The agent observes the stock market data in the form of candlestick price charts, as shown in the following image for your reference:

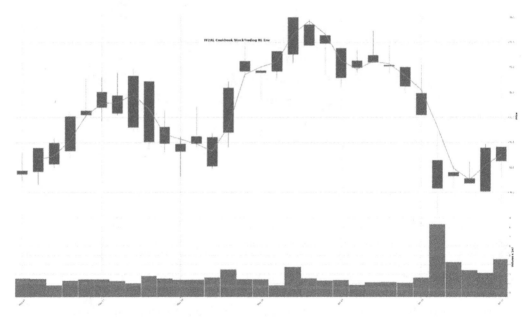

Figure 5.2 – Visual observation of StockTradingVisualContinuousEnvironment

The agent's action and learning progress can be seen in the following image, which was generated by the `render()` method:

Figure 5.3 – Visualization of the live account balance and the trade actions taken by the agent in the current time window

That concludes this recipe and this chapter. Happy training!

6
Reinforcement Learning in the Real World – Building Intelligent Agents to Complete Your To-Dos

An RL Agent needs to interact with the environment to learn and train. Training RL Agents for real-world applications usually comes with physical limitations and challenges. This is because the Agent could potentially cause damage to the real-world system it is dealing with while learning. Fortunately, there are a lot of tasks in the real world that do not necessarily have such challenges, and yet can be very useful for completing the day-to-day real-world tasks that are available in our To-Do lists!

The recipes in this chapter will help you build RL Agents that can complete tasks on the internet, ranging from responding to annoying popups, booking flights on the web, managing emails and social media accounts, and more. We can do all of this without using a bunch of APIs that change over time or utilizing hardcoded scripts that stop working when a web page is updated. You will be training the Agents to complete such To-Do tasks by using the mouse and keyboard, just like how a human would! This chapter will also help you build the **WebGym** API, which is an OpenAI Gym-compatible generic RL learning environment interface that you can use to convert more than 50+ web tasks into training environments for RL and train your own RL Agents.

Specifically, the following recipes will be covered in this chapter:

- Building learning environments for real-world RL
- Building an RL Agent to complete tasks on the web – Call to Action
- Building a visual auto-login bot
- Training an RL Agent to automate flight booking for your travel
- Training an RL Agent to manage your emails
- Training an RL Agent to automate your social media account management

Let's get started!

Technical requirements

The code in this book has been extensively tested on Ubuntu 18.04 and Ubuntu 20.04, which means it should work with later versions of Ubuntu if Python 3.6+ is available. With Python 3.6+ installed, along with the necessary Python packages listed in the Getting ready sections of each recipe, the code should run fine on Windows and Mac OSX too. It is advised that you create and use a Python virtual environment named `tf2rl-cookbook` to install the packages and run the code in this book. Installing Miniconda or Anaconda for Python virtual environment management is recommended. You will also need to install the Chromium chrome driver on your system. On Ubuntu 18.04+, you can install it by using the `sudo apt-get install chromium-chromedriver` command.

The complete code for each recipe in each chapter will be available here: `https://github.com/PacktPublishing/Tensorflow-2-Reinforcement-Learning-Cookbook`.

Building learning environments for real-world RL

This recipe will teach you how to set up and build WebGym, a **World of Bits (WoB)**-based OpenAI Gym compatible learning platform for training RL Agents for world wide web-based real-world tasks. WoB is an open domain platform for web-based Agents. For more information about WoB, check out the following link: `http://proceedings.mlr.press/v70/shi17a/shi17a.pdf`.

WebGym provides learning environments for Agents to perceive the world wide web how we (humans) perceive it – using the pixels rendered on our display screen. The Agent interacts with the environment using keyboard and mouse events as actions. This allows the Agent to experience the world wide web how we do, which means we don't need to make any additional modifications for the Agents to train. This allows us to train RL Agents that can directly work with web-based pages and applications to complete real-world tasks.

The following image shows a sample **Click-To-Action (CTA)** environment, where the task is to click on a specific link to get to the next page or step in the process:

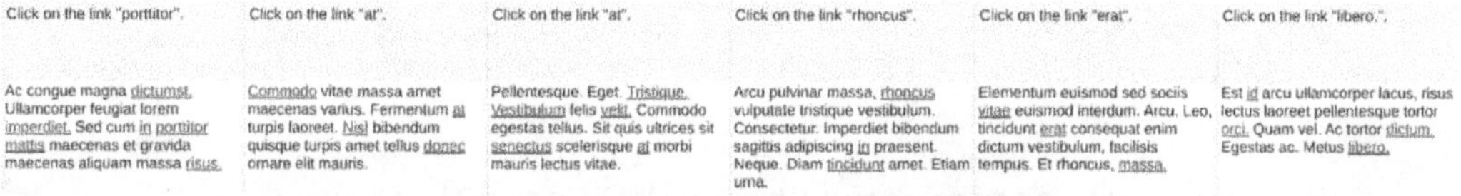

Figure 6.1 – Sample CTA task requiring a specific link to be clicked

Another example of a CTA task is depicted in the following image:

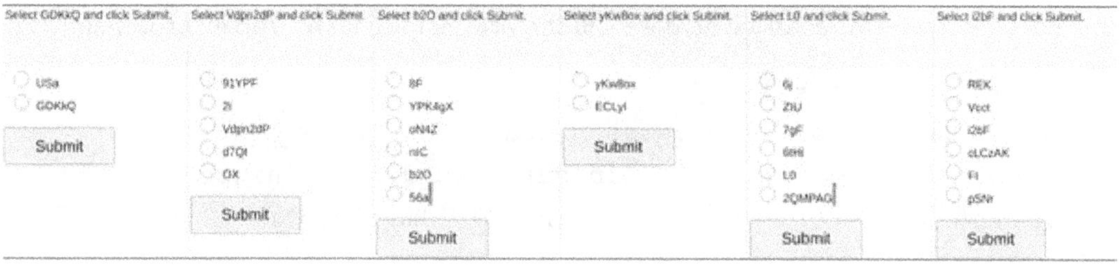

Figure 6.2 – Sample CTA task requiring a specific option to be selected and submitted

Let's get started!

Getting ready

To complete this recipe, you will need to activate the `tf2rl-cookbook` Python/conda virtual environment. Make sure that you update the environment so that it matches the latest conda environment specification file (`tfrl-cookbook.yml`) in this cookbook's code repository. WebGym is built on top of the `miniwob-plusplus` benchmark, which has also been made available as part of this book's code repository for ease of use.

Now, let's begin!

How to do it...

We will build WebGym by defining the custom `reset` and `step` methods. Then, we will define the state and action spaces for the training environments. First, we'll look at the implementation of the `miniwob_env` module. Let's get started:

1. Let's begin by importing the necessary Python modules:

   ```
   import os


   import gym
   from PIL import Image


   from miniwob.action import MiniWoBCoordClick
   from miniwob.environment import MiniWoBEnvironment
   ```

2. Let's specify the directory where we will import the local `miniwob` environment:

   ```
   cur_path_dir = \
       os.path.dirname(os.path.realpath(__file__))
   miniwob_dir = os.path.join(cur_path_dir, "miniwob",
                             "html", "miniwob")
   ```

3. Now, we can start to subclass `MiniWoBEnvironment`. We can then call the super class's initialization function to initialize the environment and set the values for `base_url` before we configure the `miniwob` environment:

   ```
   class MiniWoBEnv(MiniWoBEnvironment, gym.Env):
       def __init__(
           self,
           env_name: str,
   ```

```
        obs_im_shape,
        num_instances: int = 1,
        miniwob_dir: str = miniwob_dir,
        seeds: list = [1],
    ):
        super().__init__(env_name)
        self.base_url = f"file://{miniwob_dir}"
        self.configure(num_instances=num_instances,
                seeds=seeds, base_url=self.base_url)
        # self.set_record_screenshots(True)
        self.obs_im_shape = obs_im_shape
```

4. It's time to customize the reset (...) method. To allow environments to be randomized, we will use a seeds argument to take a random seed. This can be used to generate random start states and tasks so that the Agent we train does not overfit to a fixed/static web page:

```
def reset(self, seeds=[1], mode=None,
record_screenshots=False):
    """Forces stop and start all instances.

    Args:
        seeds (list[object]): Random seeds to set for
        each instance;
            If specified, len(seeds) must be equal to
            the number of instances.
            A None entry in the list = do not set a
            new seed.
        mode (str): If specified, set the data mode
            to this value before starting new
            episodes.
        record_screenshots (bool): Whether to record
            screenshots of the states.
    Returns:
        states (list[MiniWoBState])
    """
    miniwob_state = super().reset(seeds, mode,
```

```
                      record_screenshots=True)

    return [
        state.screenshot.resize(self.obs_im_shape,
                                Image.ANTIALIAS)
        for state in miniwob_state
    ]
```

5. Next, we will redefine the `step` (...) method. Let's complete the implementation in two steps. First, we will define the method with docstrings that explain the arguments:

```
def step(self, actions):
    """Applies an action on each instance and returns
    the results.

    Args:
        actions (list[MiniWoBAction or None])

    Returns:
        tuple (states, rewards, dones, info)
            states (list[PIL.Image.Image])
            rewards (list[float])
            dones (list[bool])
            info (dict): additional debug
            information.
                Global debug information is directly
                in the root level
                Local information for instance i is
                in info['n'][i]
    """
```

6. In this step, we will complete our implementation of the `step` (...) method:

```
states, rewards, dones, info = \
                        super().step(actions)
# Obtain screenshot & Resize image obs to match
# config
```

```
        img_states = [
            state.screenshot.resize(self.obs_im_shape) \
            if not dones[i] else None
            for i, state in enumerate(states)
        ]
        return img_states, rewards, dones, info
```

7. That completes our `MiniWoBEnv` class implementation! To test our class implementation and to understand how to use the class, we will write a quick `main()` function:

```
if __name__ == "__main__":
    env = MiniWoBVisualEnv("click-pie")
    for _ in range(10):
        obs = env.reset()
        done = False
        while not done:
            action = [MiniWoBCoordClick(90, 150)]
            obs, reward, done, info = env.step(action)
            [ob.show() for ob in obs if ob is not None]
    env.close()
```

8. You can save the preceding script as `miniwob_env.py` and execute it to see the sample environment being acted on by a random Agent. In the next few steps, we will extend `MiniWoBEnv` in order to create an OpenAI Gym-compatible learning environment interface. Let's begin by creating a new file named `envs.py` and include with the following imports:

```
import gym.spaces
import numpy as np
import string

from miniwob_env import MiniWoBEnv
from miniwob.action import MiniWoBCoordClick, MiniWoBType
```

9. For the first environment, we will implement the `MiniWoBVisualClickEnv` class:

```python
class MiniWoBVisualClickEnv(MiniWoBEnv):
    def __init__(self, name, num_instances=1):
        """RL environment with visual observations and
            touch/mouse-click action space
            Two dimensional, continuous-valued action
            space allows Agents to specify (x, y)
            coordinates on the visual rendering to click/
            touch to interact with the world-of bits

        Args:
            name (str): Name of the supported \
            MiniWoB-PlusPlus environment
            num_instances (int, optional): Number of \
            parallel env instances. Defaults to 1.
        """
        self.miniwob_env_name = name
        self.task_width = 160
        self.task_height = 210
        self.obs_im_width = 64
        self.obs_im_height = 64
        self.num_channels = 3  # RGB
        self.obs_im_size = (self.obs_im_width, \
                            self.obs_im_height)
        super().__init__(self.miniwob_env_name,
                         self.obs_im_size,
                         num_instances)
```

10. Let's also define the observation and action space for this environment in the __init__ method:

```python
        self.observation_space = gym.spaces.Box(
            0,
            255,
            (self.obs_im_width, self.obs_im_height,
            self.num_channels),
```

```
        dtype=int,
    )
    self.action_space = gym.spaces.Box(
        low=np.array([0, 0]),
        high=np.array([self.task_width,
                        self.task_height]),
        shape=(2,),
        dtype=int,
    )
```

11. Next, we will further extend the `reset (...)` method to provide an OpenAI Gym-compatible interface method:

```
def reset(self, seeds=[1]):
    """Forces stop and start all instances.

    Args:
        seeds (list[object]): Random seeds to set for
        each instance;
            If specified, len(seeds) must be equal to
            the number of instances.
            A None entry in the list = do not set a
            new seed.
    Returns:
        states (list[PIL.Image])
    """
    obs = super().reset(seeds)
    # Click somewhere to Start!
    # miniwob_state, _, _, _ = super().step(
    # self.num_instances * [MiniWoBCoordClick(10,10)]
    # )
    return obs
```

12. The next important piece is the `step` method. We will implement it in the following two steps:

```
def step(self, actions):
    """Applies an action on each instance and returns
        the results.

    Args:
        actions (list[(x, y) or None]);
            - x is the number of pixels from the left
              of browser window
            - y is the number of pixels from the top of
              browser window

    Returns:
        tuple (states, rewards, dones, info)
            states (list[PIL.Image.Image])
            rewards (list[float])
            dones (list[bool])
            info (dict): additional debug
            information.
                Global debug information is directly
                in the root level
                Local information for instance i is
                in info['n'][i]
    """
```

13. To complete the `step` method's implementation, let's check if the dimensions of the actions are as expected and then bind the actions if necessary. Finally, we must execute a step in the environment:

```
assert (
    len(actions) == self.num_instances
), f"Expected len(actions)={self.num_instances}.\
    Got {len(actions)}."

def clamp(action, low=self.action_space.low,\
          high=self.action_space.high):
```

```
            low_x, low_y = low
            high_x, high_y = high
            return (
                max(low_x, min(action[0], high_x)),
                max(low_y, min(action[1], high_y)),
            )
        miniwob_actions = \
            [MiniWoBCoordClick(*clamp(action)) if action\
            is not None else None for action in actions]
        return super().step(miniwob_actions)
```

14. We can use a descriptive name for the class to register the environment with the Gym registry:

```
class MiniWoBClickButtonVisualEnv(MiniWoBVisualClickEnv):
    def __init__(self, num_instances=1):
        super().__init__("click-button", num_instances)
```

15. Finally, to register the environment with OpenAI Gym's registry locally, we must add the environment registration information to the __init__.py file:

```
import sys
import os

from gym.envs.registration import register

sys.path.append(os.path.dirname(os.path.abspath(__
file__)))

_AVAILABLE_ENVS = {
    "MiniWoBClickButtonVisualEnv-v0": {
        "entry_point": \
            "webgym.envs:MiniWoBClickButtonVisualEnv",
        "discription": "Click the button on a web page",
    }
}
for env_id, val in _AVAILABLE_ENVS.items():
```

```
register(id=env_id,
         entry_point=val.get("entry_point"))
```

With that, we have completed this recipe!

How it works...

We have extended the implementation of `MiniWoB-plusplus` in `MiniWoBEnv` so that we can use file-based web pages to represent tasks. We extended the `MiniWoBEnv` class even further to provide an OpenAI Gym-compatible interface in `MiniWoBVisualClickEnv`.

To get a clear picture of how an RL Agent will be learning to complete the task in this environment, consider the following screenshot. Here, the Agent tries to understand the objective of the task by trying out different actions, which in this environment translates to clicking on different areas of the web page (represented on the right-hand side by blue dots). Eventually, the RL Agent clicks on the correct button and starts to understand what the task description means, as well as what the buttons are intended for, since it was rewarded for clicking on the correct spot:

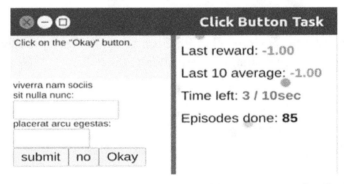

Figure 6.3 – Visualizing the Agent's actions while it's learning to complete the CTA task

Now, it's time to move on to the next recipe!

Building an RL Agent to complete tasks on the web – Call to Action

This recipe will teach you how to implement an RL training script so that you can train an RL Agent to handle **Call-To-Action** (**CTA**) type tasks for you. CTA buttons are the actionable buttons that you typically find on web pages that you need to click in order to proceed to the next step. While there are several CTA button examples available, some common examples include the OK/Cancel dialog boxes, where you need you to click to acknowledge/dismiss the pop-up notification, and the Click to learn more button. In this recipe, you will instantiate a RL training environment that provides visual rendering for the web pages containing a CTA task. You will be training a **proximal policy optimization** (**PPO**)-based deep RL Agent that's been implemented using TensorFlow 2.x to learn how to complete the task at hand.

The following image illustrates a set of observations from a randomized CTA environment (with different seeds) so that you understand the task that the Agent will be solving:

Figure 6.4 – Screenshot of the Agent's observations from a randomized CTA environment

Let's begin!

Getting ready

To complete this recipe, you will need to activate the tf2rl-cookbook Python/conda virtual environment. Make sure that you update the environment so that it matches the latest conda environment specification file (tfrl-cookbook.yml) in this cookbook's code repository. If the following import statements run without any issues, then you are ready to get started:

```
import argparse
import os
from datetime import datetime
import gym
import numpy as np
import tensorflow as tf
from tensorflow.keras.layers import
```

```
(Conv2D,Dense,Dropout,Flatten,Input,Lambda,MaxPool2D,)
import webgym  # Used to register webgym environments
```

Let's begin!

How to do it...

In this recipe, we will be implementing a complete training script, including command-line argument parsing for training hyperparameter configuration. As you may have noticed from the `import` statements, we will be using Keras's functional API for TensorFlow 2.x to implement the **deep neural networks (DNNs)** we will be using as part of the Agent's algorithm implementation.

The following steps will guide you through the implementation:

1. Let's begin by defining the command-line arguments for the CTA Agent training script:

```
parser = argparse.ArgumentParser(prog="TFRL-Cookbook-Ch5-
Click-To-Action-Agent")
parser.add_argument("--env",
default="MiniWoBClickButtonVisualEnv-v0")
parser.add_argument("--update-freq", type=int,
default=16)
parser.add_argument("--epochs", type=int, default=3)
parser.add_argument("--actor-lr", type=float,
default=1e-4)
parser.add_argument("--Critic-lr", type=float,
default=1e-4)
parser.add_argument("--clip-ratio", type=float,
default=0.1)
parser.add_argument("--gae-lambda", type=float,
default=0.95)
parser.add_argument("--gamma", type=float, default=0.99)
parser.add_argument("--logdir", default="logs")
```

2. Next, we will create a TensorBoard logger so that we can log and visualize the live training progress of the CTA Agent:

```
args = parser.parse_args()
logdir = os.path.join(
    args.logdir, parser.prog, args.env, \
    datetime.now().strftime("%Y%m%d-%H%M%S")
)
print(f"Saving training logs to:{logdir}")
writer = tf.summary.create_file_writer(logdir)
```

3. In the following steps, we will implement the `Actor` class. However, we will begin by implementing the `__init__` method:

```
class Actor:
    def __init__(self, state_dim, action_dim,
    action_bound, std_bound):
        self.state_dim = state_dim
        self.action_dim = action_dim
        self.action_bound = np.array(action_bound)
        self.std_bound = std_bound
        self.weight_initializer = \
            tf.keras.initializers.he_normal()
        self.eps = 1e-5
        self.model = self.nn_model()
        self.model.summary()  # Print a summary of the
        @ Actor model
        self.opt = \
            tf.keras.optimizers.Nadam(args.actor_lr)
```

4. Next, we will define the DNN that will represent the Actor's model. We will split the implementation of the DNN into multiple steps as it's going to be a bit long due to several neural network layers being stacked together. As the first and main processing step, we will implement a block by stacking convolution-pooling-convolution-pooling layers:

```
    def nn_model(self):
        obs_input = Input(self.state_dim)
```

```
conv1 = Conv2D(
    filters=64,
    kernel_size=(3, 3),
    strides=(1, 1),
    padding="same",
    input_shape=self.state_dim,
    data_format="channels_last",
    activation="relu",
)(obs_input)
pool1 = MaxPool2D(pool_size=(3, 3), strides=1)\
                    (conv1)
conv2 = Conv2D(
    filters=32,
    kernel_size=(3, 3),
    strides=(1, 1),
    padding="valid",
    activation="relu",
)(pool1)
pool2 = MaxPool2D(pool_size=(3, 3), strides=1)\
                    (conv2)
```

5. Now, we will flatten the output from the pooling layer so that we can start using fully connected or dense layers with dropout to generate the output we expect from the Actor network:

```
flat = Flatten()(pool2)
dense1 = Dense(
    16, activation="relu", \
        kernel_initializer=self.weight_initializer
)(flat)
dropout1 = Dropout(0.3)(dense1)
dense2 = Dense(
    8, activation="relu", \
        kernel_initializer=self.weight_initializer
)(dropout1)
dropout2 = Dropout(0.3)(dense2)
# action_dim[0] = 2
```

```
output_val = Dense(
    self.action_dim[0],
    activation="relu",
    kernel_initializer=self.weight_initializer,
) (dropout2)
```

6. We need to scale and clip the predicted value so that the values are bounded and lie within the range we expect the actions to be in. Let's use the **Lambda layer** to implement custom clipping and scaling, as shown in the following code snippet:

```
mu_output = Lambda(
    lambda x: tf.clip_by_value(x * \
        self.action_bound, 1e-9, self.action_bound)
) (output_val)
std_output_1 = Dense(
    self.action_dim[0],
    activation="softplus",
    kernel_initializer=self.weight_initializer,
) (dropout2)
std_output = Lambda(
    lambda x: tf.clip_by_value(
        x * self.action_bound, 1e-9, \
        self.action_bound / 2
    )
) (std_output_1)
return tf.keras.models.Model(
    inputs=obs_input, outputs=[mu_output, std_
output], name="Actor"
)
```

7. That completes our nn_model implementation. Now, let's define a convenience function to get an action, given a state:

```
def get_action(self, state):
    # Convert [Image] to np.array(np.adarray)
    state_np = np.array([np.array(s) for s in state])
    if len(state_np.shape) == 3:
        # Convert (w, h, c) to (1, w, h, c)
```

```
        state_np = np.expand_dims(state_np, 0)
    mu, std = self.model.predict(state_np)
    action = np.random.normal(mu, std + self.eps, \
                        size=self.action_dim).astype(
        "int"
    )
    # Clip action to be between 0 and max obs screen
    # size
    action = np.clip(action, 0, self.action_bound)
    # 1 Action per instance of env; Env expects:
    # (num_instances, actions)
    action = (action,)
    log_policy = self.log_pdf(mu, std, action)
    return log_policy, action
```

8. Now, it's time to implement the main train method. This will update the parameters of the Actor network:

```
    def train(self, log_old_policy, states, actions,
    gaes):
        with tf.GradientTape() as tape:
            mu, std = self.model(states, training=True)
            log_new_policy = self.log_pdf(mu, std,
                                        actions)
            loss = self.compute_loss(log_old_policy,
                        log_new_policy, actions, gaes)
        grads = tape.gradient(loss,
                        self.model.trainable_variables)
        self.opt.apply_gradients(zip(grads,
                        self.model.trainable_variables))
        return loss
```

9. Although we are using `compute_loss` and `log_pdf` in the preceding `train` method, we haven't really defined them yet! Let's implement them one after the other, starting with the `compute_loss` method:

```python
def compute_loss(self, log_old_policy,
log_new_policy, actions, gaes):
    # Avoid INF in exp by setting 80 as the upper
    # bound since,
    # tf.exp(x) for x>88 yeilds NaN (float32)
    ratio = tf.exp(
        tf.minimum(log_new_policy - \
            tf.stop_gradient(log_old_policy), 80)
    )
    gaes = tf.stop_gradient(gaes)
    clipped_ratio = tf.clip_by_value(
        ratio, 1.0 - args.clip_ratio, 1.0 + \
        args.clip_ratio
    )
    surrogate = -tf.minimum(ratio * gaes, \
                            clipped_ratio * gaes)
    return tf.reduce_mean(surrogate)
```

10. In this step, we will implement the `log_pdf` method:

```python
def log_pdf(self, mu, std, action):
    std = tf.clip_by_value(std, self.std_bound[0],
                            self.std_bound[1])
    var = std ** 2
    log_policy_pdf = -0.5 * (action - mu) ** 2 / var\
                     - 0.5 * tf.math.log(
        var * 2 * np.pi
    )
    return tf.reduce_sum(log_policy_pdf, 1,
                        keepdims=True)
```

11. The previous step concludes out Actor implementation. Now, it's time to start implementing the `Critic` class:

```
class Critic:
    def __init__(self, state_dim):
        self.state_dim = state_dim
        self.weight_initializer = \
            tf.keras.initializers.he_normal()
        self.model = self.nn_model()
        self.model.summary()  # Print a summary of the
        # Critic model
        self.opt = \
            tf.keras.optimizers.Nadam(args.Critic_lr)
```

12. Next up is the `Critic` class's neural network model. Like the Actor's neural network model, this is going to be a DNN. We will split the implementation into a few steps. First, let's implement a convolution-pooling-convolution-pooling block:

```
obs_input = Input(self.state_dim)
conv1 = Conv2D(
    filters=64,
    kernel_size=(3, 3),
    strides=(1, 1),
    padding="same",
    input_shape=self.state_dim,
    data_format="channels_last",
    activation="relu",
) (obs_input)
pool1 = MaxPool2D(pool_size=(3, 3), strides=2)\
                    (conv1)
conv2 = Conv2D(
    filters=32,
    kernel_size=(3, 3),
    strides=(1, 1),
    padding="valid",
    activation="relu",
) (pool1)
```

```
pool2 = MaxPool2D(pool_size=(3, 3), strides=2)\
        (conv2)
```

13. While we could stack more blocks or layers to deepen the neural network, for our current task, we already have a sufficient number of parameters in the DNN to learn how to perform well at the CTA task. Let's add the fully connected layers so that we can eventually produce the state-conditioned action value:

```
flat = Flatten()(pool2)
dense1 = Dense(
    16, activation="relu", \
        kernel_initializer=self.weight_initializer
)(flat)
dropout1 = Dropout(0.3)(dense1)
dense2 = Dense(
    8, activation="relu", \
        kernel_initializer=self.weight_initializer
)(dropout1)
dropout2 = Dropout(0.3)(dense2)
value = Dense(
    1, activation="linear", \
        kernel_initializer=self.weight_initializer
)(dropout2)
```

14. Let's implement a method that will compute the Critic's learning loss, which is essentially the mean-squared error between the temporal difference learning targets and the values predicted by the Critic:

```
def compute_loss(self, v_pred, td_targets):
    mse = tf.keras.losses.MeanSquaredError()
    return mse(td_targets, v_pred)
```

15. Let's finalize our `Critic` class by implementing the `train` method to update the Critic's parameters:

```
def train(self, states, td_targets):
    with tf.GradientTape() as tape:
```

```
        v_pred = self.model(states, training=True)
        # assert v_pred.shape == td_targets.shape
        loss = self.compute_loss(v_pred, \
                            tf.stop_gradient(td_targets))
    grads = tape.gradient(loss, \
                    self.model.trainable_variables)
    self.opt.apply_gradients(zip(grads, \
                    self.model.trainable_variables))
    return loss
```

16. Now, we can utilize the Actor and the Critic implementation to build our PPO Agent so that it can work with high-dimensional (image) observations. Let's begin by defining the PPOAgent class's __init__ method:

```
class PPOAgent:
    def __init__(self, env):
        self.env = env
        self.state_dim = self.env.observation_space.shape
        self.action_dim = self.env.action_space.shape
        # Set action_bounds to be within the actual
        # task-window/browser-view of the Agent
        self.action_bound = [self.env.task_width,
                                self.env.task_height]
        self.std_bound = [1e-2, 1.0]

        self.actor = Actor(
            self.state_dim, self.action_dim,
            self.action_bound, self.std_bound
        )
        self.Critic = Critic(self.state_dim)
```

17. We will be using **Generalized Advantage Estimates** (GAE) to update our policy. So, let's implement a method that will calculate the GAE target values:

```
    def gae_target(self, rewards, v_values, next_v_value,
    done):
        n_step_targets = np.zeros_like(rewards)
```

```
        gae = np.zeros_like(rewards)
        gae_cumulative = 0
        forward_val = 0

        if not done:
            forward_val = next_v_value

        for k in reversed(range(0, len(rewards))):
            delta = rewards[k] + args.gamma * \
                    forward_val - v_values[k]
            gae_cumulative = args.gamma * \
                                args.gae_lambda * \
                                gae_cumulative + delta
            gae[k] = gae_cumulative
            forward_val = v_values[k]
            n_step_targets[k] = gae[k] + v_values[k]
        return gae, n_step_targets
```

18. We are at the core of this script! Let's define the training routine for the deep PPO Agent. We will split the implementation into multiple steps to make it easy to follow. We will begin with the outermost loop, which must be running for a configurable maximum number of episodes:

```
    def train(self, max_episodes=1000):
        with writer.as_default():
            for ep in range(max_episodes):
                state_batch = []
                action_batch = []
                reward_batch = []
                old_policy_batch = []

                episode_reward, done = 0, False

                state = self.env.reset()
                prev_state = state
                step_num = 0
```

19. Next, we will implement the logic for stepping through the environment and handling the end of an episode by checking the done values from the environments:

```
while not done:
    log_old_policy, action = \
        self.actor.get_action(state)
    next_state, reward, dones, _ = \
        self.env.step(action)
    step_num += 1
    print(
        f"ep#:{ep} step#:{step_num} \
        step_rew:{reward} \
        action:{action} dones:{dones}"
    )
    done = np.all(dones)
    if done:
        next_state = prev_state
    else:
        prev_state = next_state
    state = np.array([np.array(s) for s\
                    in state])
    next_state = np.array([np.array(s) \
                    for s in next_state])
    reward = np.reshape(reward, [1, 1])
    log_old_policy = np.reshape(
                    log_old_policy, [1, 1])
    state_batch.append(state)
    action_batch.append(action)
    reward_batch.append((reward + 8) / 8)
    old_policy_batch.append(
                    log_old_policy)
```

20. Next, we will implement the logic that will check for the end of an episode or if it is time to update and perform an update step:

```
if len(state_batch) >= \
args.update_freq or done:
    states = \
        np.array([state.squeeze() \
        for state in state_batch])
    # Convert ([x, y],) to [x, y]
    actions = np.array([action[0] \
        for action in action_batch])
    rewards = np.array(
        [reward.squeeze() for reward\
        in reward_batch]
    )
    old_policies = np.array(
        [old_pi.squeeze() for old_pi\
        in old_policy_batch]
    )
    v_values = self.Critic.model.\
        predict(states)
    next_v_value = self.Critic.\
        model.predict(next_state)
    gaes, td_targets = \
    self.gae_target(
        rewards, v_values, \
        next_v_value, done
    )
    actor_losses, Critic_losses=[],[]
```

21. Now that we have the updated GAE targets, we can train the Actor and Critic networks and log the losses and other training metrics for tracking purposes:

```
for epoch in range(args.epochs):
    actor_loss = \
        self.actor.train(
            old_policies, states,
            actions, gaes
```

```
                                )
                    actor_losses.append(
                         actor_loss)
                Critic_loss = \
                      self.Critic.train(states,
                                    td_targets)
                Critic_losses.append(
                                  Critic_loss)
           # Plot mean actor & Critic losses
           # on every update
           tf.summary.scalar("actor_loss",
                 np.mean(actor_losses),
                 step=ep)
           tf.summary.scalar(
                "Critic_loss",
                 np.mean(Critic_losses),
                 step=ep
           )
           state_batch = []
           action_batch = []
           reward_batch = []
           old_policy_batch = []
        episode_reward += reward[0][0]
        state = next_state[0]
```

22. Finally, let's implement the __main__ function to train the CTA Agent:

```
if __name__ == "__main__":
    env_name = "MiniWoBClickButtonVisualEnv-v0"
    env = gym.make(env_name)
    cta_Agent = PPOAgent(env)
    cta_Agent.train()
```

That completes this recipe! Let's briefly recap on how it works.

How it works...

In this recipe, we implemented a PPO-based deep RL Agent and provided a training mechanism to develop a CTA Agent. Note that for simplicity, we used one instance of the environment, though the code can scale for a greater number of environment instances to speed up training.

To understand how the Agent training progresses, consider the following sequence of images. During the initial stages of training, when the Agent is trying to understand the task and the objective of the task, the Agent may just be executing random actions (exploration) or even clicking outside the screen, as shown in the following screenshot:

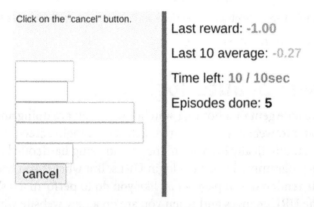

Figure 6.5 – Agent clicking outside the screen (no visible blue dot) during initial exploration

As the Agent learns by stumbling upon the correct button to click, it starts to make progress. The following screenshot shows the Agent making some progress:

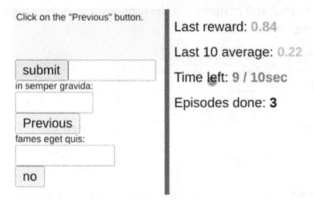

Figure 6.6 – Deep PPO Agent making progress in the CTA task

Finally, when the episode is complete or ends (due to a time limit), the Agent receives an observation similar to the one shown in the following screenshot (left):

Figure 6.7 – End of episode observation (left) and summary of performance (right)

Now, it's time to move on to the next recipe!

Building a visual auto-login bot

Imagine that you have an Agent or a bot that watches what you are doing and automatically logs you into websites whenever you click on a login screen. While browser plugins exist that can automatically log you in, they do so using hardcoded scripts that only work on the pre-programmed website's login URLs. But what if you had an Agent that only relied on the rendered web page – just like you do to perform a task – and worked even when the URL changes and when you are on a new website with no prior saved data? How cool would that be?! This recipe will help you develop a script that will train an Agent to log in on a web page! You will learn how to randomize, customize, and increase the generality of the Agent to get it to work on any login screen.

An example of randomizing and customizing the usernames and passwords for a task can be seen in the following image:

Figure 6.8 – Sample observations from a randomized user login task

Let's get started!

Getting ready

To complete this recipe, make sure you have the latest version. First, you will need to activate the tf2rl-cookbook Python/conda virtual environment. Make sure that you update the environment so that it matches the latest conda environment specification file (tfrl-cookbook.yml) in this cookbook's code repository. If the following import statements run without any issues, then you are ready to get started:

```
import argparse
import os
from datetime import datetime
import gym
import numpy as np
import tensorflow as tf
from tensorflow.keras.layers import
(Conv2D,Dense,Dropout,Flatten,Input,Lambda,MaxPool2D,)
import webgym  # Used to register webgym environments
```

Let's begin!

How to do it...

In this recipe, we will implement the deep RL-based Login Agent using the PPO algorithm.

Let's get started:

1. First, let's set up the training script's command-line arguments and logging:

    ```
    parser = argparse.ArgumentParser(prog="TFRL-Cookbook-Ch5-
    Login-Agent")
    parser.add_argument("--env",
    default="MiniWoBLoginUserVisualEnv-v0")
    parser.add_argument("--update-freq", type=int,
    default=16)
    parser.add_argument("--epochs", type=int, default=3)
    parser.add_argument("--actor-lr", type=float,
    default=1e-4)
    parser.add_argument("--Critic-lr", type=float,
    default=1e-4)
    parser.add_argument("--clip-ratio", type=float,
    default=0.1)
    ```

```
parser.add_argument("--gae-lambda", type=float,
default=0.95)
parser.add_argument("--gamma", type=float, default=0.99)
parser.add_argument("--logdir", default="logs")

args = parser.parse_args()
logdir = os.path.join(
    args.logdir, parser.prog, args.env, \
    datetime.now().strftime("%Y%m%d-%H%M%S")
)
print(f"Saving training logs to:{logdir}")
writer = tf.summary.create_file_writer(logdir)
```

2. We can now directly jump into the `Critic` class's definition:

```
class Critic:
    def __init__(self, state_dim):
        self.state_dim = state_dim
        self.weight_initializer = \
            tf.keras.initializers.he_normal()
        self.model = self.nn_model()
        self.model.summary()  # Print a summary of the
        # Critic model
        self.opt = \
            tf.keras.optimizers.Nadam(args.Critic_lr)
```

3. Now, let's define the DNN for the Critic model. We'll begin by implementing a perception block composed of convolution-pooling-convolution-pooling. In the subsequent steps, we'll add more depth to the network by stacking another perception block:

```
    def nn_model(self):
        obs_input = Input(self.state_dim)
        conv1 = Conv2D(
            filters=64,
            kernel_size=(3, 3),
            strides=(1, 1),
            padding="same",
```

```
            input_shape=self.state_dim,
            data_format="channels_last",
            activation="relu",
        )(obs_input)
        pool1 = MaxPool2D(pool_size=(3, 3), strides=2)\
                            (conv1)
        conv2 = Conv2D(
            filters=32,
            kernel_size=(3, 3),
            strides=(1, 1),
            padding="valid",
            activation="relu",
        )(pool1)
        pool2 = MaxPool2D(pool_size=(3, 3), strides=2)
                            (conv2)
```

4. Next, we will add another perception block so that we can extract more features:

```
        conv3 = Conv2D(
            filters=16,
            kernel_size=(3, 3),
            strides=(1, 1),
            padding="valid",
            activation="relu",
        )(pool2)
        pool3 = MaxPool2D(pool_size=(3, 3), strides=1)\
                            (conv3)
        conv4 = Conv2D(
            filters=8,
            kernel_size=(3, 3),
            strides=(1, 1),
            padding="valid",
            activation="relu",
        )(pool3)
        pool4 = MaxPool2D(pool_size=(3, 3), strides=1)\
    (conv4)
```

5. Next, we will add a flattening layer, followed by fully connected (dense) layers, to bring down the shape of the network's output to a single action value:

```
flat = Flatten()(pool4)
dense1 = Dense(
    16, activation="relu",
    kernel_initializer=self.weight_initializer
)(flat)
dropout1 = Dropout(0.3)(dense1)
dense2 = Dense(
    8, activation="relu",
    kernel_initializer=self.weight_initializer
)(dropout1)
dropout2 = Dropout(0.3)(dense2)
value = Dense(
    1, activation="linear",
    kernel_initializer=self.weight_initializer
)(dropout2)

return tf.keras.models.Model(inputs=obs_input,
                outputs=value, name="Critic")
```

6. To finalize our Critic implementation, let's define the `compute_loss` method and the `update` method in order to train the parameters:

```
def compute_loss(self, v_pred, td_targets):
    mse = tf.keras.losses.MeanSquaredError()
    return mse(td_targets, v_pred)


def train(self, states, td_targets):
    with tf.GradientTape() as tape:
        v_pred = self.model(states, training=True)
        # assert v_pred.shape == td_targets.shape
        loss = self.compute_loss(v_pred,
                    tf.stop_gradient(td_targets))
    grads = tape.gradient(loss,
                self.model.trainable_variables)
```

```
        self.opt.apply_gradients(zip(grads,
                            self.model.trainable_variables))
    return loss
```

7. We can now work on implementing the `Actor` class. We'll initialize the `Actor` class in this step and continue our implementation in the subsequent steps:

```
class Actor:
    def __init__(self, state_dim, action_dim,
    action_bound, std_bound):
        self.state_dim = state_dim
        self.action_dim = action_dim
        self.action_bound = np.array(action_bound)
        self.std_bound = std_bound
        self.weight_initializer = \
            tf.keras.initializers.he_normal()
        self.eps = 1e-5
        self.model = self.nn_model()
        self.model.summary()  # Print a summary of the
        # Actor model
        self.opt = tf.keras.optimizers.Nadam(
                                    args.actor_lr)
```

8. We will use a similar DNN architecture for our Actor as we did in our Critic implementation. So, the `nn_model` method's implementation will remain the same except for the last few layers, where the Actor and Critic's implementation will vary. The Actor network model produces the mean and the standard deviation as output. This depends on the action space dimensions. On the other hand, the Critic network produces a state-conditioned action value, irrespective of the dimensions of the action space. The layers that differ from the Critic's DNN implementation are listed here:

```
# action_dim[0] = 2
output_val = Dense(
    self.action_dim[0],
    activation="relu",
    kernel_initializer=self.weight_initializer,
```

```
) (dropout2)
# Scale & clip x[i] to be in range [0,
# action_bound[i]]
mu_output = Lambda(
    lambda x: tf.clip_by_value(x * \
        self.action_bound, 1e-9, self.action_bound)
) (output_val)
std_output_1 = Dense(
    self.action_dim[0],
    activation="softplus",
    kernel_initializer=self.weight_initializer,
) (dropout2)
std_output = Lambda(
    lambda x: tf.clip_by_value(
        x * self.action_bound, 1e-9,
        self.action_bound / 2
    )
) (std_output_1)
return tf.keras.models.Model(
    inputs=obs_input, outputs=[mu_output,
        std_output], name="Actor"
)
```

9. Let's implement some methods that will compute the Actor's loss and `log_pdf`:

```
def log_pdf(self, mu, std, action):
    std = tf.clip_by_value(std, self.std_bound[0],
                            self.std_bound[1])
    var = std ** 2
    log_policy_pdf = -0.5 * (action - mu) ** 2 / var\
                        - 0.5 * tf.math.log(
        var * 2 * np.pi
    )
    return tf.reduce_sum(log_policy_pdf, 1,
                            keepdims=True)
```

```python
def compute_loss(self, log_old_policy,
                 log_new_policy, actions, gaes):
    # Avoid INF in exp by setting 80 as the upper
    # bound since,
    # tf.exp(x) for x>88 yeilds NaN (float32)
    ratio = tf.exp(
        tf.minimum(log_new_policy - \
                   tf.stop_gradient(log_old_policy), 80)
    )
    gaes = tf.stop_gradient(gaes)
    clipped_ratio = tf.clip_by_value(
        ratio, 1.0 - args.clip_ratio, 1.0 + \
        args.clip_ratio
    )
    surrogate = -tf.minimum(ratio * gaes,
                            clipped_ratio * gaes)
    return tf.reduce_mean(surrogate)
```

10. With the help of these helper methods, our training method implementation becomes simpler:

```python
def train(self, log_old_policy, states, actions,
          gaes):
    with tf.GradientTape() as tape:
        mu, std = self.model(states, training=True)
        log_new_policy = self.log_pdf(mu, std,
                                      actions)
        loss = self.compute_loss(log_old_policy,
                                  log_new_policy,
                                  actions, gaes)
    grads = tape.gradient(loss,
                          self.model.trainable_variables)
    self.opt.apply_gradients(zip(grads,
                             self.model.trainable_variables))
    return loss
```

11. Finally, let's implement a method that will get an action from the Actor when it's given a state as input:

```python
def get_action(self, state):
    # Convert [Image] to np.array(np.adarray)
    state_np = np.array([np.array(s) for s in state])
    if len(state_np.shape) == 3:
        # Convert (w, h, c) to (1, w, h, c)
        state_np = np.expand_dims(state_np, 0)
    mu, std = self.model.predict(state_np)
    action = np.random.normal(mu, std + self.eps,
                        size=self.action_dim).astype(
        "int"
    )
    # Clip action to be between 0 and max obs
    # screen size
    action = np.clip(action, 0, self.action_bound)
    # 1 Action per instance of env; Env expects:
    # (num_instances, actions)
    action = (action,)
    log_policy = self.log_pdf(mu, std, action)
    return log_policy, action
```

12. That completes our Actor implementation. We can now tie both the Actor and the Critic together using the PPOAgent class implementation. Since the GAE target calculations were discussed in the previous recipe, we will skip this and focus on the training method's implementation:

```python
while not done:
    # self.env.render()
    log_old_policy, action = \
        self.actor.get_action(state)
    next_state, reward, dones, _ = \
        self.env.step(action)
    step_num += 1
    # Convert action[2] from int idx to
    # char for verbose printing
```

```
action_print = []
for a in action:  # Map apply
    action_verbose = (a[:2], \
    self.get_typed_char(a[2]))
    action_print.append(
                    action_verbose)
print(
    f"ep#:{ep} step#:{step_num}
    step_rew:{reward} \
    action:{action_print} \
    dones:{dones}"
)
done = np.all(dones)
if done:
    next_state = prev_state
else:
    prev_state = next_state
state = np.array([np.array(s) for \
    s in state])
next_state = np.array([np.array(s) \
    for s in next_state])
reward = np.reshape(reward, [1, 1])
log_old_policy = np.reshape(
                log_old_policy, [1, 1])
state_batch.append(state)
action_batch.append(action)
reward_batch.append((reward + 8) / 8)
old_policy_batch.append(\
                log_old_policy)
```

13. The Agent's update is performed at a preset frequency in terms of the number of samples collected or at the end of every episode – whichever occurs first:

```
if len(state_batch) >= \
    args.update_freq or done:
    states = np.array([state.\
                squeeze() for state\
```

```
                              in state_batch])
        actions = np.array([action[0]\
                for action in action_batch])
        rewards = np.array(
                [reward.squeeze() for reward\
                in reward_batch])
        old_policies = np.array(
                [old_pi.squeeze() for old_pi\
                in old_policy_batch])
        v_values = self.Critic.model.\
                        predict(states)
        next_v_value = self.Critic.\
                model.predict(next_state)


        gaes, td_targets = \
                    self.gae_target(
                rewards, v_values, \
                next_v_value, done)
        actor_losses, Critic_losses=[],[]
        for epoch in range(args.epochs):
            actor_loss = \
                self.actor.train(
                    old_policies, states,
                    actions, gaes)
            actor_losses.append(
                            actor_loss)
            Critic_loss = self.Critic.\
                train(states, td_targets)
            Critic_losses.append(
                            Critic_loss)
```

14. Finally, we can run `MiniWoBLoginUserVisualEnv-v0` and train the Agent using the following snippet of code:

```
if __name__ == "__main__":
    env_name = "MiniWoBLoginUserVisualEnv-v0"
    env = gym.make(env_name)
    cta_Agent = PPOAgent(env)
    cta_Agent.train()
```

That completes our script for the auto-login Agent. It's time for you to run the script to see the Agent's training process in action!

How it works...

The login task involves clicking on the correct form field and typing in the correct username and/or password. For an Agent to be able to do this, it needs to master how to use a mouse and keyboard, in addition to processing the visual web page to understand the task and the web login form. With enough samples, the deep RL Agent will learn a policy to complete this task. Let's take a look at the state of the Agent's progress, snapshotted at different stages.

The following image shows the Agent successfully entering the username and correctly clicking on the password field to enter the password, but not being able to complete the task yet:

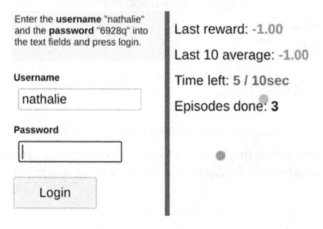

Figure 6.9 – Screenshot of a trained Agent successfully entering the username but not a password

In the following image, you can see that the Agent has learned to enter both the username and password, but they are not quite right for the task to be classed as complete:

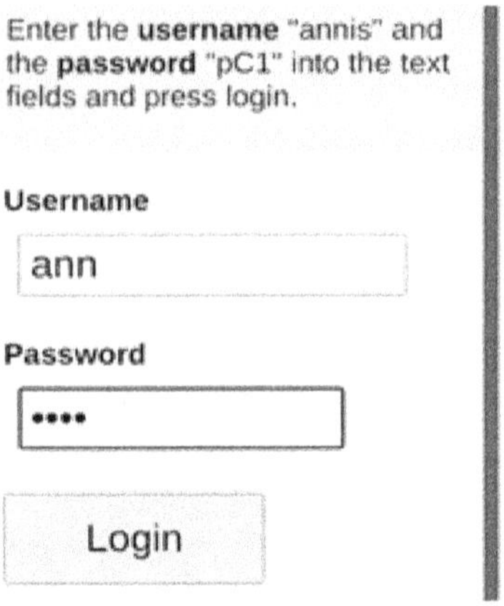

Figure 6.10 – Agent entering both the username and password but incorrectly

The same Agent with a different checkpoint, after several thousand more episodes of learning, is close to completing the task, as shown in the following image

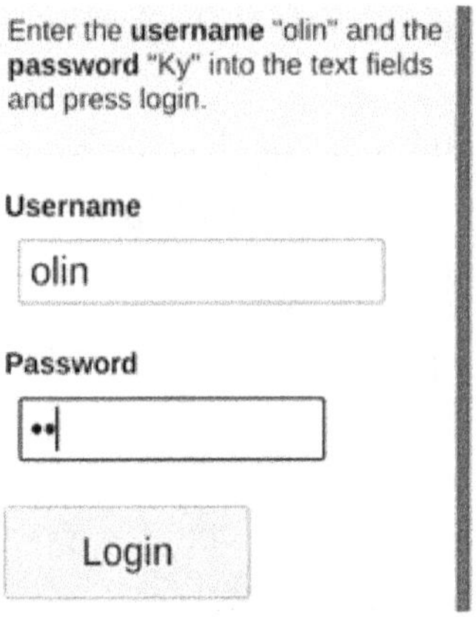

Figure 6.11 – A well-trained Agent model about to complete the login task successfully

Now that you understand how the Agent works and behaves, you can customize it to your liking and use use cases to train the Agent to automatically log into any custom website you want!

Training an RL Agent to automate flight booking for your travel

In this recipe, you will learn how to implement a deep RL Agent based on the **Deep Deterministic Policy Gradient (DDPG)** algorithm using TensorFlow 2.x and train the Agent to visually operate flight booking websites using a keyboard and mouse to book flights! This task is quite useful but complicated due to the varying amount of task parameters we need to implement, such as source city, destination, date, and more. The following image shows a sample of the start states from a randomized `MiniWoBBookFlightVisualEnv` flight booking environment:

Figure 6.12 – Sample start-state observations from the randomized MiniWoBBookFlightVisualEnv environment

Let's get started!

Getting ready

To complete this recipe, you will need to activate the `tf2rl-cookbook` Python/conda virtual environment. Make sure that you update the environment so that it matches the latest conda environment specification file (`tfrl-cookbook.yml`) in this cookbook's code repository. If the following `import` statements run without any issues, then you are ready to get started:

```
import argparse
import os
import random
from collections import import deque
from datetime import datetime

import gym
import numpy as np
import tensorflow as tf
from tensorflow.keras.layers import (Conv2D, Dense, Dropout,
```

```
Flatten, Input,Lambda,MaxPool2D)
import webgym  # Used to register webgym environments
```

How to do it...

In this recipe, we will be implementing a complete training script that you will be able to customize and train to book flights!

Let's get started:

1. First, let's expose the hyperparameters as configurable arguments to the training script:

    ```
    parser = argparse.ArgumentParser(
        prog="TFRL-Cookbook-Ch5-SocialMedia-Mute-User-
    DDPGAgent"
    )
    parser.add_argument("--env", default="Pendulum-v0")
    parser.add_argument("--actor_lr", type=float,
    default=0.0005)
    parser.add_argument("--Critic_lr", type=float,
    default=0.001)
    parser.add_argument("--batch_size", type=int, default=64)
    parser.add_argument("--tau", type=float, default=0.05)
    parser.add_argument("--gamma", type=float, default=0.99)
    parser.add_argument("--train_start", type=int,
    default=2000)
    parser.add_argument("--logdir", default="logs")
    args = parser.parse_args()
    ```

2. Next, we'll set up TensorBoard logging for live visualization of the training progress:

    ```
    logdir = os.path.join(
        args.logdir, parser.prog, args.env, \
        datetime.now().strftime("%Y%m%d-%H%M%S")
    )
    print(f"Saving training logs to:{logdir}")
    writer = tf.summary.create_file_writer(logdir)
    ```

3. We'll be using a Replay Buffer to implement Experience Reply. Let's implement a simple `ReplayBuffer` class:

```python
class ReplayBuffer:
    def __init__(self, capacity=10000):
        self.buffer = deque(maxlen=capacity)

    def store(self, state, action, reward, next_state,
    done):
        self.buffer.append([state, action, reward,
                            next_state, done])

    def sample(self):
        sample = random.sample(self.buffer,
                               args.batch_size)
        states, actions, rewards, next_states, done = \
                        map(np.asarray, zip(*sample))
        states = \
            np.array(states).reshape(args.batch_size, -1)
        next_states = np.array(next_states).\
                    reshape(args.batch_size, -1)
        return states, actions, rewards, next_states,\
        done

    def size(self):
        return len(self.buffer)
```

4. Let's start by implementing the `Actor` class:

```python
class Actor:
    def __init__(self, state_dim, action_dim,
    action_bound):
        self.state_dim = state_dim
        self.action_dim = action_dim
        self.action_bound = action_bound
        self.weight_initializer = \
            tf.keras.initializers.he_normal()
        self.eps = 1e-5
```

```
self.model = self.nn_model()
self.opt = tf.keras.optimizers.Adam(
                                args.actor_lr)
```

5. The DNN model for the Actor will be composed of two perception blocks, each containing convolution-pooling-convolution-pooling layers, as in our previous recipe. We'll skip this here and look at the implementation of the `train` method instead. The full source code, as always, will be available in this cookbook's code repository. Let's continue with our `train` and `predict` method implementations:

```
def train(self, states, q_grads):
    with tf.GradientTape() as tape:
        grads = tape.gradient(
            self.model(states),
            self.model.trainable_variables,
            -q_grads
        )
    self.opt.apply_gradients(zip(grads, \
                    self.model.trainable_variables))

def predict(self, state):
    return self.model.predict(state)
```

6. The last piece of our `Actor` class is to implement a function to get the action:

```
def get_action(self, state):
    # Convert [Image] to np.array(np.adarray)
    state_np = np.array([np.array(s) for s in state])
    if len(state_np.shape) == 3:
        # Convert (w, h, c) to (1, w, h, c)
        state_np = np.expand_dims(state_np, 0)
    action = self.model.predict(state_np)
    # Clip action to be between 0 and max obs
    # screen size
    action = np.clip(action, 0, self.action_bound)
    # 1 Action per instance of env; Env expects:
    # (num_instances, actions)
    return action
```

7. With that, our `Actor` class is ready. Now, we can move on and implement the `Critic` class:

```
class Critic:
    def __init__(self, state_dim, action_dim):
        self.state_dim = state_dim
        self.action_dim = action_dim
        self.weight_initializer = \
            tf.keras.initializers.he_normal()
        self.model = self.nn_model()
        self.opt = \
            tf.keras.optimizers.Adam(args.Critic_lr)
```

8. Similar to the `Actor` class's DNN model, we will be reusing a similar architecture for our `Critic` class from the previous recipe, with two perception blocks. You can refer to the full source code of this recipe or the DNN implementation in the previous recipe for completeness. Let's jump into the implementation of the `predict` and `g_gradients` computations for the Q function:

```
    def predict(self, inputs):
        return self.model.predict(inputs)

    def q_gradients(self, states, actions):
        actions = tf.convert_to_tensor(actions)
        with tf.GradientTape() as tape:
            tape.watch(actions)
            q_values = self.model([states, actions])
            q_values = tf.squeeze(q_values)
        return tape.gradient(q_values, actions)
```

9. In order to update our Critic model, we need a loss to drive the parameter updates and an actual training step to perform the update. In this step, we will implement these two core methods:

```
    def compute_loss(self, v_pred, td_targets):
        mse = tf.keras.losses.MeanSquaredError()
        return mse(td_targets, v_pred)

    def train(self, states, actions, td_targets):
```

```
        with tf.GradientTape() as tape:
            v_pred = self.model([states, actions],
                                    training=True)
            assert v_pred.shape == td_targets.shape
            loss = self.compute_loss(v_pred,
                        tf.stop_gradient(td_targets))
        grads = tape.gradient(loss,
                    self.model.trainable_variables)
        self.opt.apply_gradients(zip(grads,
                    self.model.trainable_variables))
        return loss
```

10. It's time to bring the Actor and Critic together to implement the DDPGAgent! Let's dive into it:

```
class DDPGAgent:
    def __init__(self, env):
        self.env = env
        self.state_dim = self.env.observation_space.shape
        self.action_dim = self.env.action_space.shape
        self.action_bound = self.env.action_space.high
        self.buffer = ReplayBuffer()
        self.actor = Actor(self.state_dim,
                    self.action_dim, self.action_bound)
        self.Critic = Critic(self.state_dim,
                            self.action_dim)
        self.target_actor = Actor(self.state_dim,
                    self.action_dim, self.action_bound)
        self.target_Critic = Critic(self.state_dim,
                            self.action_dim)
        actor_weights = self.actor.model.get_weights()
        Critic_weights = self.Critic.model.get_weights()
        self.target_actor.model.set_weights(
                            actor_weights)
        self.target_Critic.model.set_weights
                            (Critic_weights)
```

11. Let's implement a method that will update the target models of our Actor and Critic:

```
def update_target(self):
    actor_weights = self.actor.model.get_weights()
    t_actor_weights = \
        self.target_actor.model.get_weights()
    Critic_weights = self.Critic.model.get_weights()
    t_Critic_weights = \
        self.target_Critic.model.get_weights()
    for i in range(len(actor_weights)):
        t_actor_weights[i] = (args.tau * \
        actor_weights[i] + (1 - args.tau) * \
        t_actor_weights[i])
    for i in range(len(Critic_weights)):
        t_Critic_weights[i] = (args.tau * \
        Critic_weights[i] + (1 - args.tau) * \
        t_Critic_weights[i])
    self.target_actor.model.set_weights(
                        t_actor_weights)
    self.target_Critic.model.set_weights(
                        t_Critic_weights)
```

12. Next, we will implement a method that will compute the temporal difference targets:

```
def get_td_target(self, rewards, q_values, dones):
    targets = np.asarray(q_values)
    for i in range(q_values.shape[0]):
        if dones[i]:
            targets[i] = rewards[i]
        else:
            targets[i] = args.gamma * q_values[i]
    return targets
```

13. Because we are using a deterministic policy gradient and a policy without a distribution to sample from, we will be using a noise function to sample around the action predicted by the Actor network. The **Ornstein Uhlenbeck (OU)** noise process is a popular choice for DDPG Agents. We'll implement this here:

```
def add_ou_noise(self, x, rho=0.15, mu=0, dt=1e-1,
                    sigma=0.2, dim=1):
    return (
        x + rho * (mu - x) * dt + sigma * \
        np.sqrt(dt) * np.random.normal(size=dim))
```

14. Next, we will implement a method that will replay the experience from the replay buffer:

```
def replay_experience(self):
    for _ in range(10):
        states, actions, rewards, next_states, \
            dones = self.buffer.sample()
        target_q_values = self.target_Critic.predict(
            [next_states,
             self.target_actor.predict(next_states)])
        td_targets = self.get_td_target(rewards,
                            target_q_values, dones)

        self.Critic.train(states, actions,
                            td_targets)

        s_actions = self.actor.predict(states)
        s_grads = self.Critic.q_gradients(states,
                                            s_actions)
        grads = np.array(s_grads).reshape(
                            (-1, self.action_dim))
        self.actor.train(states, grads)
        self.update_target()
```

15. The last but most crucial thing we must do in our Agent implementation is implement the `train` method. We will split the implementation into a few steps. First, we will start with the outermost loop, which must run for a maximum number of episodes:

```python
def train(self, max_episodes=1000):
    with writer.as_default():
        for ep in range(max_episodes):
            step_num, episode_reward, done = 0, 0,\
                                                False

            state = self.env.reset()
            prev_state = state
            bg_noise = np.random.randint(
                self.env.action_space.low,
                self.env.action_space.high,
                self.env.action_space.shape,
            )
```

16. Next, we will implement the inner loop, which will run until the end of an episode:

```python
while not done:
    # self.env.render()
    action = self.actor.get_action(state)
    noise = self.add_ou_noise(bg_noise,\
                            dim=self.action_dim)
    action = np.clip(action + noise, 0, \
            self.action_bound).astype("int")

    next_state, reward, dones, _ = \
                        self.env.step(action)
    done = np.all(dones)
    if done:
        next_state = prev_state
    else:
        prev_state = next_state
```

17. We are not done yet! We still need to update our Replay Buffer with the new experience that the Agent has collected:

```
for (s, a, r, s_n, d) in zip(next_state,\
action, reward, next_state, dones):
    self.buffer.store(s, a, \
                (r + 8) / 8, s_n, d)
    episode_reward += r
step_num += 1  # 1 across
# num_instances
print(f"ep#:{ep} step#:{step_num} \
    step_rew:{reward} \
    action:{action} dones:{dones}")
bg_noise = noise
state = next_state
```

18. Are we done?! Almost! We just have to remember to replay the experience when the buffer size is bigger than the batch size we used for training:

```
if (self.buffer.size() >= args.batch_size
    and self.buffer.size() >= \
    args.train_start):
    self.replay_experience()
print(f"Episode#{ep} \
        Reward:{episode_reward}")
tf.summary.scalar("episode_reward", \
                episode_reward, \
                step=ep)
```

19. That completes our implementation. Now, we can launch the Agent training on the Visual Flight Booking environment using the following __main__ function:

```
if __name__ == "__main__":

    env_name = "MiniWoBBookFlightVisualEnv-v0"
    env = gym.make(env_name)
    Agent = DDPGAgent(env)
    Agent.train()
```

That's it!

How it works...

The DDPG Agent collects a series of samples from the flight booking environment as it explores and uses this experience to update its policy parameters through the Actor and Critic updates. The OU noise we discussed earlier allows the Agent to explore while using a deterministic action policy. The flight booking environment is quite complex as it requires the Agent to master both the keyboard and the mouse, in addition to understanding the task by looking at visual images of the task description (visual text parsing), inferring the intended task objective, and executing the actions in the correct sequence. The following screenshot shows the performance of the Agent upon completing a sufficiently large number of episodes of training:

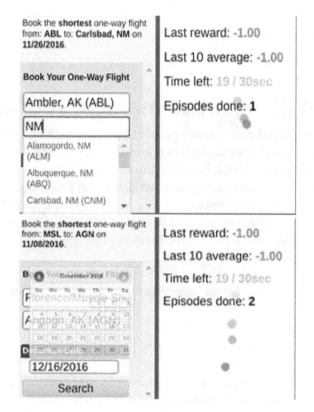

Figure 6.13 – A screenshot of the Agent performing the flight booking task at different stages of learning

The following screenshot shows the Agent's screen after the Agent progressed to the final stage of the task (although it's not close to completing the task):

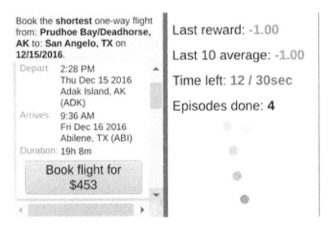

Figure 6.14 – Screenshot of the Agent progressing all the way to the final stage of the flight booking task

With that, we will move on to the next recipe!

Training an RL Agent to manage your emails

Email has become an integral part of many people's lives. The number of emails that an average working professional goes through in a workday is growing daily. While a lot of email filters exist for spam control, how nice would it be to have an intelligent Agent that can perform a series of email management tasks that just provide a task description (through text or speech via speech-to-text) and are not limited by any APIs that have rate limits? In this recipe, you will develop a deep RL Agent and train it on email management tasks! A set of sample tasks can be seen in the following image:

Figure 6.15 – A sample set of observations from the randomized
MiniWoBEmailInboxImportantVisualEnv environment

Let's get into the details!

Getting ready

To complete this recipe, make sure you have the latest version. First, you will need to activate the tf2rl-cookbook Python/conda virtual environment. Make sure that you update the environment so that it matches the latest conda environment specification file (tfrl-cookbook.yml) in this cookbook's code repository. If the following import statements run without any issues, then you are ready to get started:

```
import tensorflow as tf
from tensorflow.keras.layers import (
    Conv2D,
    Dense,
    Dropout,
    Flatten,
    Input,
    Lambda,
    MaxPool2D,
)
import webgym  # Used to register webgym environments
```

Let's begin!

How to do it...

Follow these steps to implement a deep RL Agent and train it to manage important emails:

1. First, we will define an ArgumentParser so that we can configure the script from the command line. For a complete list of configurable hyperparameters, please refer to the source code for this recipe:

```
parser = argparse.ArgumentParser(
    prog="TFRL-Cookbook-Ch5-Important-Emails-Manager-
Agent"
)
parser.add_argument("--env",
default="MiniWoBEmailInboxImportantVisualEnv-v0")
```

2. Next, let's set up TensorBoard logging:

```
args = parser.parse_args()
logdir = os.path.join(
```

```
        args.logdir, parser.prog, args.env, \
        datetime.now().strftime("%Y%m%d-%H%M%S")
)
print(f"Saving training logs to:{logdir}")
writer = tf.summary.create_file_writer(logdir)
```

3. Now, we can initialize the Actor class:

```
class Actor:
    def __init__(self, state_dim, action_dim,
    action_bound, std_bound):
        self.state_dim = state_dim
        self.action_dim = action_dim
        self.action_bound = np.array(action_bound)
        self.std_bound = std_bound
        self.weight_initializer = \
            tf.keras.initializers.he_normal()
        self.eps = 1e-5
        self.model = self.nn_model()
        self.model.summary()  # Print a summary of the ·
        # Actor model
        self.opt = \
            tf.keras.optimizers.Nadam(args.actor_lr)
```

4. Because the observations in our email management environment are visual (images), we will need perception capabilities for the Actor in our Agent. For this, we must make use of convolution-based perception blocks, as follows:

```
    def nn_model(self):
        obs_input = Input(self.state_dim)
        conv1 = Conv2D(
            filters=32,
            kernel_size=(3, 3),
            strides=(1, 1),
            padding="same",
            input_shape=self.state_dim,
```

```
        data_format="channels_last",
        activation="relu",
    )(obs_input)
    pool1 = MaxPool2D(pool_size=(3, 3), strides=1)\
                    (conv1)
    conv2 = Conv2D(
        filters=32,
        kernel_size=(3, 3),
        strides=(1, 1),
        padding="valid",
        activation="relu",
    )(pool1)
    pool2 = MaxPool2D(pool_size=(3, 3), strides=1)\
                    (conv2)
```

5. Now, let's add more perception blocks comprising convolutions, followed by max pooling layers:

```
    conv3 = Conv2D(
        filters=16,
        kernel_size=(3, 3),
        strides=(1, 1),
        padding="valid",
        activation="relu",
    )(pool2)
    pool3 = MaxPool2D(pool_size=(3, 3), strides=1)\
                    (conv3)
    conv4 = Conv2D(
        filters=16,
        kernel_size=(3, 3),
        strides=(1, 1),
        padding="valid",
        activation="relu",
    )(pool3)
    pool4 = MaxPool2D(pool_size=(3, 3), strides=1)\
                    (conv4)
```

6. Now, we are ready to flatten the DNN output to produce the mean (mu) and standard deviation that we want as the output from the Actor. First, let's add the flattening layer and the dense layers:

```
flat = Flatten()(pool4)
dense1 = Dense(
    16, activation="relu", \
        kernel_initializer=self.weight_initializer
)(flat)
dropout1 = Dropout(0.3)(dense1)
dense2 = Dense(
    8, activation="relu", \
        kernel_initializer=self.weight_initializer
)(dropout1)
dropout2 = Dropout(0.3)(dense2)
# action_dim[0] = 2
output_val = Dense(
    self.action_dim[0],
    activation="relu",
    kernel_initializer=self.weight_initializer,
)(dropout2)
```

7. We are now ready to define the final layers of our Actor network. These will helps us produce mu and std, as we discussed in the previous step:

```
# Scale & clip x[i] to be in range [0,
                                action_bound[i]]
mu_output = Lambda(
    lambda x: tf.clip_by_value(x * \
        self.action_bound, 1e-9, self.action_bound)
)(output_val)
std_output_1 = Dense(
    self.action_dim[0],
    activation="softplus",
    kernel_initializer=self.weight_initializer,
)(dropout2)
std_output = Lambda(
```

```
        lambda x: tf.clip_by_value(
            x * self.action_bound, 1e-9, \
            self.action_bound / 2))(std_output_1)
    return tf.keras.models.Model(
        inputs=obs_input, outputs=[mu_output, \
            std_output], name="Actor"
    )
```

8. That completes our Actor's DNN model implementation. To implement the remaining methods to complete the Actor class, please refer to the full code for this recipe, which can be found in this cookbook's code repository. We will now focus on defining the interfaces for the `Critic` class:

```
class Critic:
    def __init__(self, state_dim):
        self.state_dim = state_dim
        self.weight_initializer = \
                tf.keras.initializers.he_normal()
        self.model = self.nn_model()
        self.model.summary()  # Print a summary of the
        # Critic model
        self.opt = \
                tf.keras.optimizers.Nadam(args.critic_lr)
```

9. The Critic's DNN model is also based on the same convolutional neural network architecture that we used for the `Actor`. For completeness, please refer to this recipe's full source code, which is available in this cookbook's code repository. We will implement the loss computation and training method here:

```
    def compute_loss(self, v_pred, td_targets):
        mse = tf.keras.losses.MeanSquaredError()
        return mse(td_targets, v_pred)

    def train(self, states, td_targets):
        with tf.GradientTape() as tape:
            v_pred = self.model(states, training=True)
```

```
        # assert v_pred.shape == td_targets.shape
        loss = self.compute_loss(v_pred, \
                    tf.stop_gradient(td_targets))
    grads = tape.gradient(loss, \
                    self.model.trainable_variables)
    self.opt.apply_gradients(zip(grads, \
                    self.model.trainable_variables))
    return loss
```

10. With that, we can now define our Agent's class:

```
class PPOAgent:
    def __init__(self, env):
        self.env = env
        self.state_dim = self.env.observation_space.shape
        self.action_dim = self.env.action_space.shape
        # Set action_bounds to be within the actual
        # task-window/browser-view of the Agent
        self.action_bound = [self.env.task_width, \
                            self.env.task_height]
        self.std_bound = [1e-2, 1.0]

        self.actor = Actor(
            self.state_dim, self.action_dim, \
            self.action_bound, self.std_bound
        )
        self.Critic = Critic(self.state_dim)
```

11. The preceding code should be familiar to you from the previous Agent implementations in this chapter. You can complete the remaining methods (and the training loop) based on our previous implementations. If you get stuck, you can refer to the full source code for this recipe by going to this cookbook's code repository. We will now write the `__main__` function so that we can train the Agent in `MiniWoBEmailInboxImportantVisualEnv`. This will allow us to see the Agent's learning process in action:

```python
if __name__ == "__main__":
    env_name = "MiniWoBEmailInboxImportantVisualEnv-v0"
    env = gym.make(env_name)
    cta_Agent = PPOAgent(env)
    cta_Agent.train()
```

How it works...

The PPO Agent uses convolutional neural network layers to process the high-dimensional visual inputs in the Actor and Critic classes. The PPO algorithm updates the Agent's policy parameters using a surrogate loss function that prevents the policy parameters from being drastically updated. It then keeps the policy updates within the trust region, which makes it robust to hyperparameter choices and a few other factors that may lead to instability during the Agent's training regime. The email management environment poses as a nice sequential decision-making problem for the deep RL Agent. First, the Agent has to choose the correct email from a series of emails in an inbox and then perform the desired action (starring the email and so on). The Agent only has access to the visual rendering of the inbox, so it needs to extract the task specification details, interpret the task specification, and then plan and execute the actions!

The following is a screenshot of the Agent's performance at different stages of learning (loaded from different checkpoints):

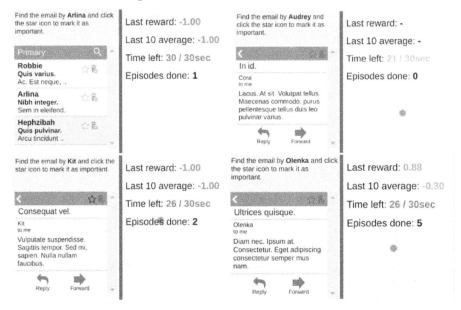

Figure 6.16 – A series of screenshots showing the Agent's learning progress

Now, let's move on to the next recipe!

Training an RL Agent to automate your social media account management

By the end of this recipe, you will have built a complete deep RL Agent training script that can be trained to perform management tasks on your social media account!

The following image shows a series of (randomized) tasks from the environment that we will be training the Agent in:

Figure 6.17 – A sample set of social media account management tasks that the Agent has been asked to solve

Note that there is a scroll bar in this task that the Agent needs to learn how to use! The tweet that's relevant to this task may be hidden from the visible part of the screen, so the Agent will have to actively explore (by sliding the scroll bar up/down) in order to progress!

Getting ready

To complete this recipe, you will need to activate the `tf2rl-cookbook` Python/conda virtual environment. Make sure that you update the environment so that it matches the latest conda environment specification file (`tfrl-cookbook.yml`) in this cookbook's code repository. If the following `import` statements run without any issues, then you are ready to get started:

```
import tensorflow as tf
from tensorflow.keras.layers import (Conv2D, Dense, Dropout,
Flatten, Input, Lambda, MaxPool2D, concatenate,)
import webgym  # Used to register webgym environments
```

How to do it...

Let's start by configuring the Agent training script. After that, you will be shown how to complete the implementation.

Let's get started:

1. Let's jump right into the implementation! We will begin with our `ReplayBuffer` implementation:

    ```
    class ReplayBuffer:
        def __init__(self, capacity=10000):
            self.buffer = deque(maxlen=capacity)

        def store(self, state, action, reward, next_state,
        done):
            self.buffer.append([state, action, reward,
                                next_state, done])

        def sample(self):
            sample = random.sample(self.buffer,
                                   args.batch_size)
    ```

```
            states, actions, rewards, next_states, done = \
                                map(np.asarray, zip(*sample))
        states = \
            np.array(states).reshape(args.batch_size, -1)
        next_states = np.array(next_states).\
                            reshape(args.batch_size, -1)
        return states, actions, rewards, next_states,\
        done

    def size(self):
        return len(self.buffer)
```

2. Next, we will implement our `Actor` class:

```
class Actor:
    def __init__(self, state_dim, action_dim,
    action_bound):
        self.state_dim = state_dim
        self.action_dim = action_dim
        self.action_bound = action_bound
        self.weight_initializer = \
            tf.keras.initializers.he_normal()
        self.eps = 1e-5
        self.model = self.nn_model()
        self.opt = \
            tf.keras.optimizers.Adam(args.actor_lr)
```

3. The next core piece is the DNN definition for our Actor:

```
    def nn_model(self):
        obs_input = Input(self.state_dim)
        conv1 = Conv2D(filters=64, kernel_size=(3, 3),\
                    strides=(1, 1), padding="same", \
                    input_shape=self.state_dim, \
                    data_format="channels_last", \
                    activation="relu")(obs_input)
        pool1 = MaxPool2D(pool_size=(3, 3), \
```

```
                        strides=1)(conv1)
conv2 = Conv2D(filters=32, kernel_size=(3, 3),\
                 strides=(1, 1), padding="valid", \
                 activation="relu",)(pool1)
pool2 = MaxPool2D(pool_size=(3, 3), strides=1)\
                 (conv2)
```

4. Depending on the complexity of the task, we can modify (add/reduce) the depth of the DNN. We will start by connecting the output of the pooling layers to the fully connected layers with dropout:

```
flat = Flatten()(pool2)
dense1 = Dense(
    16, activation="relu", \
    kernel_initializer=self.weight_initializer)\
    (flat)
dropout1 = Dropout(0.3)(dense1)
dense2 = Dense(8, activation="relu", \
    kernel_initializer=self.weight_initializer)\
    (dropout1)
dropout2 = Dropout(0.3)(dense2)
# action_dim[0] = 2
output_val = Dense(self.action_dim[0],
                   activation="relu",
                   kernel_initializer= \
                        self.weight_initializer,)\
                   (dropout2)
# Scale & clip x[i] to be in range
# [0, action_bound[i]]
mu_output=Lambda(lambda x: tf.clip_by_value(x *\
                  self.action_bound, 1e-9, \
                  self.action_bound))(output_val)
return tf.keras.models.Model(inputs=obs_input,
                  outputs=mu_output,
                  name="Actor")
```

5.   That completes our DNN model implementation for the Actor. Now, let's implement the methods that will train the Actor and get the predictions from the Actor model:

```
def train(self, states, q_grads):
    with tf.GradientTape() as tape:
        grads = tape.gradient(self.model(states),\
                          self.model.trainable_variables,\
                          -q_grads)
    self.opt.apply_gradients(zip(grads, \
        self.model.trainable_variables))


def predict(self, state):
    return self.model.predict(state)
```

6.   We can now get the actions from our Actor:

```
def get_action(self, state):
    # Convert [Image] to np.array(np.adarray)
    state_np = np.array([np.array(s) for s in state])
    if len(state_np.shape) == 3:
        # Convert (w, h, c) to (1, w, h, c)
        state_np = np.expand_dims(state_np, 0)
    action = self.model.predict(state_np)
    action = np.clip(action, 0, self.action_bound)
    return action
```

7. Let's get started with our Critic implementation. Here, we will need to implement the Agent class that we are after:

```
class Critic:
    def __init__(self, state_dim, action_dim):
        self.state_dim = state_dim
        self.action_dim = action_dim
        self.weight_initializer = \
            tf.keras.initializers.he_normal()
        self.model = self.nn_model()
        self.opt = \
            tf.keras.optimizers.Adam(args.Critic_lr)
```

8. Note that the Critic's model is initialized with `self.nn_model()`. Let's implement this here:

```
    def nn_model(self):
        obs_input = Input(self.state_dim)
        conv1 = Conv2D(filters=64, kernel_size=(3, 3),
                       strides=(1, 1), padding="same",
                       input_shape=self.state_dim,
                       data_format="channels_last",
                       activation="relu",)(obs_input)
        pool1 = MaxPool2D(pool_size=(3, 3), strides=2)\
                         (conv1)
        conv2 = Conv2D(filters=32, kernel_size=(3, 3),
                       strides=(1, 1), padding="valid",
                       activation="relu",)(pool1)
        pool2 = MaxPool2D(pool_size=(3, 3),
                          strides=2)(conv2)
```

9. We will complete our DNN architecture for the Critic by funneling the output through the fully connected layers with dropout. This way, we receive the necessary action values:

```
        flat = Flatten()(pool2)
        dense1 = Dense(16, activation="relu",
                       kernel_initializer= \
```

```
                              self.weight_initializer) (flat)
        dropout1 = Dropout(0.3)(dense1)
        dense2 = Dense(8, activation="relu",
                    kernel_initializer= \
                       self.weight_initializer)\
                    (dropout1)
        dropout2 = Dropout(0.3)(dense2)
        value = Dense(1, activation="linear",
                    kernel_initializer= \
                       self.weight_initializer)\
                    (dropout2)
        return tf.keras.models.Model(inputs=obs_input,
                                    outputs=value,
                                    name="Critic")
```

10. Now, let's implement the `q_gradients` and `compute_loss` methods. This should be pretty straightforward:

```
    def q_gradients(self, states, actions):
        actions = tf.convert_to_tensor(actions)
        with tf.GradientTape() as tape:
            tape.watch(actions)
            q_values = self.model([states, actions])
            q_values = tf.squeeze(q_values)
        return tape.gradient(q_values, actions)


    def compute_loss(self, v_pred, td_targets):
        mse = tf.keras.losses.MeanSquaredError()
        return mse(td_targets, v_pred)
```

11. Finally, we can complete the Critic's implementation by implementing the `predict` and `train` methods:

```
    def predict(self, inputs):
        return self.model.predict(inputs)


    def train(self, states, actions, td_targets):
```

```
        with tf.GradientTape() as tape:
            v_pred = self.model([states, actions],\
                                    training=True)
            assert v_pred.shape == td_targets.shape
            loss = self.compute_loss(v_pred, \
                            tf.stop_gradient(td_targets))
        grads = tape.gradient(loss, \
                        self.model.trainable_variables)
        self.opt.apply_gradients(zip(grads, \
                        self.model.trainable_variables))
        return loss
```

12. We can now utilize the Actor and Critic to implement our Agent:

```
class DDPGAgent:
    def __init__(self, env):
        self.env = env
        self.state_dim = self.env.observation_space.shape
        self.action_dim = self.env.action_space.shape
        self.action_bound = self.env.action_space.high
        self.buffer = ReplayBuffer()
        self.actor = Actor(self.state_dim,
                            self.action_dim,
                            self.action_bound)
        self.Critic = Critic(self.state_dim,
                            self.action_dim)
        self.target_actor = Actor(self.state_dim,
                                self.action_dim,
                                self.action_bound)
        self.target_Critic = Critic(self.state_dim,
                                self.action_dim)
        actor_weights = self.actor.model.get_weights()
        Critic_weights = self.Critic.model.get_weights()
        self.target_actor.model.set_weights(
                                    actor_weights)
        self.target_Critic.model.set_weights(
                                    Critic_weights)
```

13. Next, we will implement the `update_target` method, as per the DDPG algorithm:

```python
def update_target(self):
    actor_weights = self.actor.model.get_weights()
    t_actor_weights = \
        self.target_actor.model.get_weights()
    Critic_weights = self.Critic.model.get_weights()
    t_Critic_weights = \
        self.target_Critic.model.get_weights()
    for i in range(len(actor_weights)):
        t_actor_weights[i] = (args.tau * \
                              actor_weights[i] + \
                              (1 - args.tau) * \
                              t_actor_weights[i])

    for i in range(len(Critic_weights)):
        t_Critic_weights[i] = (args.tau * \
                               Critic_weights[i] + \
                               (1 - args.tau) * \
                               t_Critic_weights[i])

    self.target_actor.model.set_weights(
                               t_actor_weights)
    self.target_Critic.model.set_weights(
                               t_Critic_weights)
```

14. We will not look at the implementation of the `train` method here. Instead, we will start the outer loop's implementation, before completing it in the following steps:

```python
def train(self, max_episodes=1000):
    with writer.as_default():
        for ep in range(max_episodes):
            step_num, episode_reward, done = 0, 0, \
                                                False
            state = self.env.reset()
            prev_state = state
            bg_noise = np.random.randint(
```

```
                        self.env.action_space.low,
                        self.env.action_space.high,
                        self.env.action_space.shape)
```

15. The main inner loop's implementation is as follows:

```
while not done:
    action = self.actor.get_action(state)
    noise = self.add_ou_noise(bg_noise,
                    dim=self.action_dim)
    action = np.clip(action + noise, 0,
            self.action_bound).astype("int")
    next_state, reward, dones, _ = \
        self.env.step(action)
    done = np.all(dones)
    if done:
        next_state = prev_state
    else:
        prev_state = next_state
    for (s, a, r, s_n, d) in zip\
    (next_state, action, reward, \
    next_state, dones):
        self.buffer.store(s, a, \
                    (r + 8) / 8, s_n, d)
        episode_reward += r
    step_num += 1
    # 1 across num_instances
    bg_noise = noise
    state = next_state
    if (self.buffer.size() >= args.batch_size
    and self.buffer.size() >= \
        args.train_start):
    self.replay_experience()
tf.summary.scalar("episode_reward",
                    episode_reward,
                    step=ep)
```

16. That completes our training method implementation. For the implementation of the `replay_experience`, `add_ou_noise`, and `get_td_targets` methods, please refer to the full source code of this recipe, which can be found in this cookbook's code repository.

17. Let's write our `__main__` function so that we can start training the Agent in our social media environment:

```
if __name__ == "__main__":
    env_name = "MiniWoBSocialMediaMuteUserVisualEnv-v0"
    env = gym.make(env_name)
    Agent = DDPGAgent(env)
    Agent.train()
```

How it works...

Let's visually explore how a well-trained Agent progresses through social media management tasks. The following screenshot shows the Agent learning to use the scroll bar to "navigate" in this environment:

Figure 6.18 – The Agent learning to navigate using the scroll bar

Note that the task specification does not imply anything related to the scroll bar or the navigation, and that the Agent was able to explore and figure out that it needs to navigate in order to progress with the task! The following screenshot shows the Agent progressing much further by choosing the correct tweet but clicking on the wrong action; that is, Embed Tweet instead of the Mute button:

Figure 6.19 – The Agent clicking on Embed Tweet when the goal was to click on Mute

After 96 million episodes of training, the Agent was sufficiently able to solve the task. The following screenshot shows the Agent's performance on an evaluation episode (the Agent was loaded from a checkpoint)

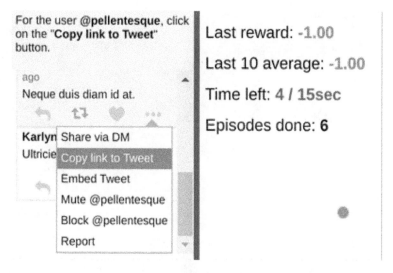

Figure 6.20 – The Agent loaded from trained parameters about to complete the task successfully

That concludes this recipe and this chapter. Happy training!

7
Deploying Deep RL Agents to the Cloud

The cloud has become the *de facto* platform of deployment for AI-based products and solutions. Deep learning models running in the cloud are becoming increasingly common. The deployment of reinforcement learning-based agents to the cloud is, however, still very limited for a variety of reasons. This chapter contains recipes to equip yourself with tools and details to get ahead of the curve and build cloud-based Simulation-as-a-Service and Agent/Bot-as-a-Service applications using deep RL.

Specifically, the following recipes are discussed in this chapter:

- Implementing the RL agent's runtime components
- Building RL environment simulators as a service
- Training RL agents using a remote simulator service
- Testing/evaluating RL agents
- Packaging RL agents for deployment – a trading bot
- Deploying RL agents to the cloud – a trading Bot-as-a-Service

Technical requirements

The code in the book is extensively tested on Ubuntu 18.04 and Ubuntu 20.04 and should work with later versions of Ubuntu if Python 3.6+ is available. With Python 3.6+ installed along with the necessary Python packages as listed before the start of each of the recipes, the code should run fine on Windows and macOS X too. It is advised to create and use a Python virtual environment named `tf2rl-cookbook` to install the packages and run the code in this book. Miniconda or Anaconda installation for Python virtual environment management is recommended.

The complete code for each recipe in each chapter is available here: `https://github.com/PacktPublishing/Tensorflow-2-Reinforcement-Learning-Cookbook`.

Implementing the RL agent's runtime components

We have looked at several agent algorithm implementations in the previous chapters. You may have noticed from recipes in the previous chapters (especially *Chapter 3, Implementing Advanced Deep RL Algorithms*), where we implemented RL agent training code, that some parts of the agent code were conditionally executed. For example, the experience replay routine was only run when a certain condition (such as the number of samples in the replay memory) was met, and so on. That begs the question: what are the essential components in an agent that is required, especially when we do not aim to train it further and only execute a learned policy?

This recipe will help you distill the implementation of the **Soft Actor-Critic** (**SAC**) agent down to the minimal set of components – those that are absolutely necessary for the runtime of your agent.

Let's get started!

Getting ready

To complete this recipe, you will first need to activate the `tf2rl-cookbook` Python/conda virtual environment. Make sure to update the environment to match the latest conda environment specification file (`tfrl-cookbook.yml`) in the cookbook's code repository. WebGym is built on top of the miniwob-plusplus benchmark (`https://github.com/stanfordnlp/miniwob-plusplus`), which is also made available as part of this book's code repository for ease of use. If the following `import` statements run without issues, you are ready to get started:

```
import functools
from collections import import deque


import numpy as np
import tensorflow as tf
import tensorflow_probability as tfp
from tensorflow.keras.layers import Concatenate, Dense, Input
from tensorflow.keras.models import Model
from tensorflow.keras.optimizers import Adam


tf.keras.backend.set_floatx("float64")
```

Now, let's begin!

How to do it...

The following steps provide the implementation details for the minimal runtime necessary to utilize an SAC agent. Let's jump right into the details:

1. First, let's implement the actor component, which is going to be a TensorFlow 2.x model:

```
def actor(state_shape, action_shape, units=(512, 256,
64)):
    state_shape_flattened = \
        functools.reduce(lambda x, y: x * y, state_shape)
    state = Input(shape=state_shape_flattened)
    x = Dense(units[0], name="L0", activation="relu")\
            (state)
    for index in range(1, len(units)):
```

```
        x = Dense(units[index], name="L{}".format(index),
                    activation="relu")(x)
    actions_mean = Dense(action_shape[0], \
                        name="Out_mean")(x)
    actions_std = Dense(action_shape[0],
                        name="Out_std")(x)
    model = Model(inputs=state, outputs=[actions_mean,
                actions_std])
    return model
```

2. Next, let's implement the critic component, which is also going to be a TensorFlow 2.x model:

```
def critic(state_shape, action_shape, units=(512, 256,
64)):
    state_shape_flattened = \
        functools.reduce(lambda x, y: x * y, state_shape)
    inputs = [Input(shape=state_shape_flattened), \
                Input(shape=action_shape)]
    concat = Concatenate(axis=-1)(inputs)
    x = Dense(units[0], name="Hidden0", \
                activation="relu")(concat)
    for index in range(1, len(units)):
        x = Dense(units[index], \
                    name="Hidden{}".format(index), \
                    activation="relu")(x)
    output = Dense(1, name="Out_QVal")(x)
    model = Model(inputs=inputs, outputs=output)
    return model
```

3. Now, let's implement a utility function to update the weights of a target model given a source TensorFlow 2.x model:

```
def update_target_weights(model, target_model,
tau=0.005):
    weights = model.get_weights()
    target_weights = target_model.get_weights()
    for i in range(len(target_weights)):
    # set tau% of target model to be new weights
        target_weights[i] = weights[i] * tau + \
                            target_weights[i] * (1 - tau)
    target_model.set_weights(target_weights)
```

4. Now, we can begin our SAC agent runtime class implementation. We will split our implementation into the following steps. Let's start with the class implementation and define the constructor arguments in this step:

```
class SAC(object):
    def __init__(
        self,
        observation_shape,
        action_space,
        lr_actor=3e-5,
        lr_critic=3e-4,
        actor_units=(64, 64),
        critic_units=(64, 64),
        auto_alpha=True,
        alpha=0.2,
        tau=0.005,
        gamma=0.99,
        batch_size=128,
        memory_cap=100000,
    ):
```

5. Let's now initialize the state/observation shapes, action shapes, and action limits/ bounds for the agent and also initialize a deque to store the agent's memory:

```
self.state_shape = observation_shape  # shape of
# observations
self.action_shape = action_space.shape  # number
# of actions
self.action_bound = \
    (action_space.high - action_space.low) / 2
self.action_shift = \
    (action_space.high + action_space.low) / 2
self.memory = deque(maxlen=int(memory_cap))
```

6. In this step, let's define and initialize the actor component:

```
# Define and initialize actor network
self.actor = actor(self.state_shape,
                   self.action_shape,
                   actor_units)
self.actor_optimizer = \
    Adam(learning_rate=lr_actor)
self.log_std_min = -20
self.log_std_max = 2
print(self.actor.summary())
```

7. Let's now define and initialize the critic components:

```
# Define and initialize critic networks
self.critic_1 = critic(self.state_shape,
                       self.action_shape,
                       critic_units)
self.critic_target_1 = critic(self.state_shape,
                              self.action_shape,
                              critic_units)
self.critic_optimizer_1 = \
    Adam(learning_rate=lr_critic)
update_target_weights(self.critic_1,
                      self.critic_target_1,
                      tau=1.0)
```

```
                self.critic_2 = critic(self.state_shape,
                                       self.action_shape,
                                       critic_units)
                self.critic_target_2 = critic(self.state_shape,
                                              self.action_shape,
                                              critic_units)
                self.critic_optimizer_2 = \
                    Adam(learning_rate=lr_critic)
                update_target_weights(self.critic_2,
                                      self.critic_target_2,
                                      tau=1.0)

                print(self.critic_1.summary())
```

8. In this step, let's initialize the temperature and target entropy for the SAC agent based on the auto_alpha flag:

```
                # Define and initialize temperature alpha and
                # target entropy
                self.auto_alpha = auto_alpha
                if auto_alpha:
                    self.target_entropy = \
                        -np.prod(self.action_shape)
                    self.log_alpha = tf.Variable(0.0,
                                                 dtype=tf.float64)
                    self.alpha = tf.Variable(0.0,
                                             dtype=tf.float64)
                    self.alpha.assign(tf.exp(self.log_alpha))
                    self.alpha_optimizer = \
                        Adam(learning_rate=lr_actor)
                else:
                    self.alpha = tf.Variable(alpha,
                                             dtype=tf.float64)
```

9. Let's complete the constructor implementation by setting the hyperparameters and initializing the training progress summary dictionary for TensorBoard logging:

```
# Set hyperparameters
self.gamma = gamma  # discount factor
self.tau = tau  # target model update
self.batch_size = batch_size


# Tensorboard
self.summaries = {}
```

10. With the constructor implementation completed, let's now move on to implement the `process_action` function, which takes the raw action from the agent and processes it so that it can be executed:

```
def process_actions(self, mean, log_std, test=False,
eps=1e-6):
    std = tf.math.exp(log_std)
    raw_actions = mean

    if not test:
        raw_actions += tf.random.normal(shape=mean.\
                        shape, dtype=tf.float64) * std

    log_prob_u = tfp.distributions.Normal(loc=mean,
                    scale=std).log_prob(raw_actions)
    actions = tf.math.tanh(raw_actions)

    log_prob = tf.reduce_sum(log_prob_u - \
                tf.math.log(1 - actions ** 2 + eps))

    actions = actions * self.action_bound + \
                self.action_shift

    return actions, log_prob
```

11. This step is crucial. We are going to implement the `act` method, which will take as input the state and generate and return the action to be executed:

```python
def act(self, state, test=False, use_random=False):
    state = state.reshape(-1)  # Flatten state
    state = np.expand_dims(state, axis=0).\
                                astype(np.float64)

    if use_random:
        a = tf.random.uniform(
            shape=(1, self.action_shape[0]),
                minval=-1, maxval=1,
                dtype=tf.float64
        )
    else:
        means, log_stds = self.actor.predict(state)
        log_stds = tf.clip_by_value(log_stds,
                                    self.log_std_min,
                                    self.log_std_max)

        a, log_prob = self.process_actions(means,
                                            log_stds,
                                            test=test)

    q1 = self.critic_1.predict([state, a])[0][0]
    q2 = self.critic_2.predict([state, a])[0][0]
    self.summaries["q_min"] = tf.math.minimum(q1, q2)
    self.summaries["q_mean"] = np.mean([q1, q2])

    return a
```

12. Finally, let's implement utility methods to load the actor and critic model weights from previously trained models:

```python
def load_actor(self, a_fn):
    self.actor.load_weights(a_fn)
    print(self.actor.summary())
```

```
def load_critic(self, c_fn):
    self.critic_1.load_weights(c_fn)
    self.critic_target_1.load_weights(c_fn)
    self.critic_2.load_weights(c_fn)
    self.critic_target_2.load_weights(c_fn)
    print(self.critic_1.summary())
```

That completes the implementation of all the necessary runtime components for the SAC RL agent!

How it works...

In this recipe, we implemented the essential runtime components for the SAC agent. The runtime components include the actor and critic model definitions, a mechanism to load weights from previously trained agent models, and an agent interface to generate actions given states using the actor's prediction and to process the prediction to generate an executable action.

The runtime components for other actor-critic-based RL agent algorithms, such as A2C, A3C, and DDPG, as well as their extensions and variants, will be very similar, if not the same.

It's now time to move on to the next recipe!

Building RL environment simulators as a service

This recipe will walk you through the process of converting your RL training environment/simulator into a service. This will allow you to offer Simulation-as-a-Service for training RL agents!

So far, we have trained several RL agents in a variety of environments using different simulators depending on the task to be solved. The training scripts used the Open AI Gym interface to talk to the environment running in the same process, or locally in a different process. This recipe will guide you through the process of converting any OpenAI Gym-compatible training environment (including your custom RL training environments) into a service that can be deployed locally or remotely as a service. Once built and deployed, an agent training client can connect to the sim server and train one or more agents remotely.

As a concrete example, we will take our `tradegym` library, which is a collection of the RL training environments for cryptocurrency and stock trading that we built in the previous chapters, and expose them through a **RESTful HTTP interface** for training RL agents.

Let's get started!

Getting ready

To complete this recipe, you will first need to activate the `tf2rl-cookbook` Python/ conda virtual environment. Make sure to update the environment to match the latest conda environment specification file (`tfrl-cookbook.yml`) in the cookbook's code repository.

We will also need to create a new Python module called `tradegym`, which contains the `crypto_trading_env.py`, `stock_trading_continuous_env.py`, `trading_utils.py`, and other custom trading environments we implemented in the previous chapters. You will find the `tradegym` module with its contents in the book's code repository as well.

How to do it...

Our implementation will contain two core modules – the `tradegym` server and the `tradegym` client, which are built based on the OpenAI Gym HTTP API. The recipe will focus on the customizations and the core components of the HTTP service interface. We will first define a minimum set of custom environments exposed as part of the `tradegym` library and then build the server and client modules:

1. Let's first make sure the minimal contents of the `tradegym` library's `__init__`. `py` file exists so that we can import these environments:

```
import sys
import os

from gym.envs.registration import register
sys.path.append(os.path.dirname(os.path.abspath(__
file__)))

_AVAILABLE_ENVS = {
    "CryptoTradingEnv-v0": {
        "entry_point": \
            "tradegym.crypto_trading_env:CryptoTradingEnv",
```

```
        "description": "Crypto Trading RL environment",
    },
    "StockTradingContinuousEnv-v0": {
        "entry_point": "tradegym.stock_trading_\
            continuous_env:StockTradingContinuousEnv",
        "description": "Stock Trading RL environment with
continous action space",
    },
}
for env_id, val in _AVAILABLE_ENVS.items():
    register(id=env_id, entry_point=val.get(
                            "entry_point"))
```

2. We can now begin our `tradegym` server implementation as `tradegym_http_server.py`. We will finalize the implementation in the following several steps. Let's begin by importing the necessary Python modules:

```
import argparse
import json
import logging
import os
import sys
import uuid
import numpy as np
import six
from flask import Flask, jsonify, request
import gym
```

3. Next, we will import the `tradegym` module to register the available environments with the Gym registry:

```
sys.path.append(os.path.dirname(os.path.abspath(__
file__)))
import tradegym  # Register tradegym envs with OpenAI Gym
# registry
```

4. Let's now look at the skeleton of the environment container class with comments describing what each method does. You can refer to the full implementation of `tradegym_http_server.py` in the book's code repository under `chapter7`. We will start with the class definition in this step and complete the skeleton in the following steps:

```
class Envs(object):
    def __init__(self):
        self.envs = {}
        self.id_len = 8   # Number of chars in instance_id
```

5. In this step, we will look at the two helper methods that are useful for managing the environment instances. They enable look up and delete operations:

```
def _lookup_env(self, instance_id):
    """Lookup environment based on instance_id and
        throw error if not found"""


def _remove_env(self, instance_id):
    """Delete environment associated with
        instance_id"""
```

6. Next, we will look at a few other methods that help with the environment management operations:

```
def create(self, env_id, seed=None):
    """Create (make) an instance of the environment
        with `env_id` and return the instance_id"""


def list_all(self):
    """Return a dictionary of all the active
        environments with instance_id as keys"""
```

```
def reset(self, instance_id):
    """Reset the environment pointed to by the
        instance_id"""


def env_close(self, instance_id):
    """Call .close() on the environment and remove
        instance_id from the list of all envs"""
```

7. The methods discussed in this step enable the core operation of the RL
    environment, which have 1-1 correspondences with the core Gym API:

```
def step(self, instance_id, action, render):
    """Perform a single step in the environment
        pointed to by the instance_id and return
        observation, reward, done and info"""


def get_action_space_contains(self, instance_id, x):
    """Check if the given environment's action space
        contains x"""


def get_action_space_info(self, instance_id):
    """Return the observation space infor for the
        given environment instance_id"""


def get_action_space_sample(self, instance_id):
    """Return a sample action for the environment
        referred by the instance_id"""


def get_observation_space_contains(self, instance_id,
j):
    """Return true is the environment's observation
        space contains `j`. False otherwise"""


def get_observation_space_info(self, instance_id):
    """Return the observation space for the
        environment referred by the instance_id"""
```

```
def _get_space_properties(self, space):
    """Return a dictionary containing the attributes
        and values of the given Gym Spce (Discrete,
        Box etc.)"""
```

8. With the preceding skeleton (and implementation), we can look at a few samples to expose these operations as REST APIs using the **Flask** Python library. We will discuss the core server application setup and the route setup for the create, reset, and step methods in the following steps. Let's look at the server application setup that exposes the endpoint handlers:

```
app = Flask(__name__)
envs = Envs()
```

9. We can now look at the REST API route definition for the **HTTP POST** endpoint at v1/envs. This takes in an env_id, which should be a valid Gym environment ID (such as our custom StockTradingContinuous-v0 or MountainCar-v0, which is available in the Gym registry) and returns an instance_id:

```
@app.route("/v1/envs/", methods=["POST"])
def env_create():
    env_id = get_required_param(request.get_json(),
                                "env_id")
    seed = get_optional_param(request.get_json(),
                              "seed", None)
    instance_id = envs.create(env_id, seed)
    return jsonify(instance_id=instance_id)
```

10. Next, we will look at the REST API route definition for the HTTP POST endpoint at v1/envs/<instance_id>/reset, where <instance_id> can be any of the IDs returned by the env_create() method:

```
@app.route("/v1/envs/<instance_id>/reset/",
           methods=["POST"])
def env_reset(instance_id):
    observation = envs.reset(instance_id)
    if np.isscalar(observation):
        observation = observation.item()
    return jsonify(observation=observation)
```

11. Next, we will look at the route definition for the endpoint at `v1/envs/<instance_id>/step`, which is the endpoint that will likely be called the most number of times in an RL training loop:

```
@app.route("/v1/envs/<instance_id>/step/",
            methods=["POST"])
def env_step(instance_id):
    json = request.get_json()
    action = get_required_param(json, "action")
    render = get_optional_param(json, "render", False)
    [obs_jsonable, reward, done, info] = envs.
step(instance_id, action, render)
    return jsonify(observation=obs_jsonable,
                reward=reward, done=done, info=info)
```

12. For the remaining route definitions on the `tradegym` server, refer to the book's code repository. We will implement a `__main__` function in the `tradegym` server script to launch the server when executed (which we will be using later in this recipe to test):

```
if __name__ == "__main__":
    parser = argparse.ArgumentParser(description="Start a
                        Gym HTTP API server")
    parser.add_argument("-l","--listen", help="interface\
                        to listen to", default="0.0.0.0")
    parser.add_argument("-p", "--port", default=6666, \
                        type=int, help="port to bind to")
    args = parser.parse_args()
    print("Server starting at: " + \
            "http://{}:{}".format(args.listen, args.port))
    app.run(host=args.listen, port=args.port, debug=True)
```

13. We will now move on to understand the implementation of the `tradegym` client. The full implementation is available in `tradegym_http_client.py` in the book's code repository under `chapter7`. We will begin by importing the necessary Python modules in this step and continue to implement the client wrapper in the following steps:

```python
import json
import logging
import os

import requests
import six.moves.urllib.parse as urlparse
```

14. The client class provides a Python wrapper to interface with the `tradegym` HTTP server. The constructor of the client class takes in the server's address (IP and port information) to connect. Let's look at the constructor implementation:

```python
class Client(object):
    def __init__(self, remote_base):
        self.remote_base = remote_base
        self.session = requests.Session()
        self.session.headers.update({"Content-type": \
                                     "application/json"})
```

15. Reproducing all the standard Gym HTTP client methods here will not be a sensible use of the space available for this book and therefore, we will focus on the core wrapper methods, such as `env_create`, `env_reset`, and `env_step`, which we will use extensively in our agent training script. For a complete implementation, please refer to the book's code repository. Let's look at the `env_create` wrapper for creating an instance of an RL simulation environment on the remote `tradegym` server:

```python
    def env_create(self, env_id):
        route = "/v1/envs/"
        data = {"env_id": env_id}
        resp = self._post_request(route, data)
        instance_id = resp["instance_id"]
        return instance_id
```

16. In this step, we will look at the wrapper method that calls the `reset` method on a specific environment using the unique `instance_id` returned by the `tradegym` server when the `env_create` call is made:

```
def env_reset(self, instance_id):
    route = "/v1/envs/{}/reset/".format(instance_id)
    resp = self._post_request(route, None)
    observation = resp["observation"]
    return observation
```

17. The most frequently used method on the `tradegym` client's `Client` class is the `step` method. Let's look at the implementation, which should look straightforward to you:

```
def env_step(self, instance_id, action,
render=False):
    route = "/v1/envs/{}/step/".format(instance_id)
    data = {"action": action, "render": render}
    resp = self._post_request(route, data)
    observation = resp["observation"]
    reward = resp["reward"]
    done = resp["done"]
    info = resp["info"]
    return [observation, reward, done, info]
```

18. With the other client wrapper methods in place, we can implement the __main__ routine to connect to the `tradegym` server and call a few methods as an example to test whether everything is working as expected. Let's write out the __main__ routine:

```
if __name__ == "__main__":
    remote_base = "http://127.0.0.1:6666"
    client = Client(remote_base)
    # Create environment
    env_id = "StockTradingContinuousEnv-v0"
    # env_id = "CartPole-v0"
    instance_id = client.env_create(env_id)
    # Check properties
    all_envs = client.env_list_all()
```

```
logger.info(f"all_envs:{all_envs}")
action_info = \
    client.env_action_space_info(instance_id)
logger.info(f"action_info:{action_info}")
obs_info = \
    client.env_observation_space_info(instance_id)
# logger.info(f"obs_info:{obs_info}")
# Run a single step
init_obs = client.env_reset(instance_id)
[observation, reward, done, info] = \
    client.env_step(instance_id, 1, True)
logger.info(f"reward:{reward} done:{done} \
                info:{info}")
```

19. We can now start to actually create a client instance and check the tradegym service! First, we need to launch the tradegym server by executing the following command:

```
(tfrl-cookbook)praveen@desktop:~/tensorflow2-
reinforcement-learning-cookbook/src/ch7-cloud-deploy-
deep-rl-agents$ python tradegym_http_server.py
```

20. Now, we can launch the tradegym client by running the following command in a separate terminal:

```
(tfrl-cookbook)praveen@desktop:~/tensorflow2-
reinforcement-learning-cookbook/src/ch7-cloud-deploy-
deep-rl-agents$ python tradegym_http_client.py
```

21. You should see an output similar to the following in the terminal where you launched the tradegym_http_client.py script:

```
all_envs:{'114c5e8f': 'StockTradingContinuousEnv-v0',
'6287385e': 'StockTradingContinuousEnv-v0', 'd55c97c0':
'StockTradingContinuousEnv-v0', 'fd355ed8':
'StockTradingContinuousEnv-v0'}
action_info:{'high': [1.0], 'low': [-1.0], 'name': 'Box',
'shape': [1]}
reward:0.0 done:False info:{}
```

That completes the recipe! Let's briefly recap on how it works.

How it works...

The `tradegym` server provides an environment container class and exposes the environment interface through a REST API. The `tradegym` client provides Python wrapper methods to interact with the RL environment through the REST API.

The `Envs` class acts as a manager for the environments instantiated on the `tradegym` server. It also acts as a container for several environments as a client can send a request to create multiple (same or different) environments. When the `tradegym` client requests the `tradegym` server using the REST API to create a new environment, the server creates an instance of the requested environment and returns a unique instance ID (example: `8kdi4289`). From that point on, the client can refer to specific environments using the instance ID. This allows the client and the agent training code to interact with multiple environments simultaneously. Thus, the `tradegym` server acts as a true service with a RESTful interface over HTTP.

Ready for the next recipe? Let's do it.

Training RL agents using a remote simulator service

In this recipe, we will look at how we can utilize a remote simulator service to train our agent. We will be reusing the SAC agent implementation from one of the previous chapters and will focus on how we can train the SAC, or any of your RL agents for that matter, using an RL simulator that is running elsewhere (on the cloud, for example) as a service. We will leverage the `tradegym` server we built in the previous recipe to provide us with the RL simulator service for this recipe.

Let's get started!

Getting ready

To complete this recipe, and to ensure that you have the latest version, you will first need to activate the `tf2rl-cookbook` Python/conda virtual environment. Make sure to update the environment to match the latest conda environment specification file (`tfrl-cookbook.yml`) in the cookbook's code repository. If the following `import` statements run without issues, you are ready to get started:

```
import datetime
import os
import sys
import logging
```

```
import gym.spaces
import numpy as np
import tensorflow as tf

sys.path.append(os.path.dirname(os.path.abspath(__file__)))
from tradegym_http_client import Client
from sac_agent_base import SAC
```

Let's get right into it.

How to do it...

We will implement the core parts of the training script and leave out command-line configuration and other non-essential functionalities to keep the script concise. Let's name our script 3_training_rl_agents_using_remote_sims.py.

Let's get started!

1. Let's first create an application-level child logger, add a stream handler to it, and then set the logging level:

```
# Create an App-level child logger
logger = logging.getLogger("TFRL-cookbook-ch7-training-
with-sim-server")
# Set handler for this logger to handle messages
logger.addHandler(logging.StreamHandler())
# Set logging-level for this logger's handler
logger.setLevel(logging.DEBUG)
```

2. Next, let's create a TensorFlow SummaryWriter to log the agent's training progress:

```
current_time = datetime.datetime.now().strftime("%Y%m%d-
%H%M%S")
train_log_dir = os.path.join("logs", "TFRL-Cookbook-Ch4-
SAC", current_time)
summary_writer = tf.summary.create_file_writer(train_log_
dir)
```

3. We can now get to the core of our implementation. Let's start the implementation of the __main__ function and continue the implementation in the following steps. Let's first set up the client to connect to the sim service using the server address:

```python
if __name__ == "__main__":

    # Set up client to connect to sim server
    sim_service_address = "http://127.0.0.1:6666"
    client = Client(sim_service_address)
```

4. Next, let's ask the server to create our desired RL training environment to train our agent:

```python
    # Set up training environment
    env_id = "StockTradingContinuousEnv-v0"
    instance_id = client.env_create(env_id)
```

5. Now, let's initialize our agent:

```python
    # Set up agent
    observation_space_info = \
        client.env_observation_space_info(instance_id)
    observation_shape = \
        observation_space_info.get("shape")
    action_space_info = \
        client.env_action_space_info(instance_id)
    action_space = gym.spaces.Box(
        np.array(action_space_info.get("low")),
        np.array(action_space_info.get("high")),
        action_space_info.get("shape"),
    )
    agent = SAC(observation_shape, action_space)
```

6. We are now ready to configure our training using a few hyperparameters:

```python
    # Configure training
    max_epochs = 30000
    random_epochs = 0.6 * max_epochs
    max_steps = 100
    save_freq = 500
```

```
reward = 0
done = False


done, use_random, episode, steps, epoch, \
episode_reward = (
    False,
    True,
    0,
    0,
    0,
    0,
)
```

7. With that, we are ready to start our outer training loop:

```
cur_state = client.env_reset(instance_id)
# Start training
while epoch < max_epochs:
    if steps > max_steps:
        done = True
```

8. Let's now handle the case where an episode has ended and done is set to True:

```
if done:
    episode += 1
    logger.info(
        f"episode:{episode} \
        cumulative_reward:{episode_reward} \
        steps:{steps} epochs:{epoch}")
    with summary_writer.as_default():
        tf.summary.scalar("Main/episode_reward",
                          episode_reward, step=episode)
        tf.summary.scalar("Main/episode_steps",
                          steps, step=episode)
    summary_writer.flush()

    done, cur_state, steps, episode_reward = (
        False,
```

```
            client.env_reset(instance_id), 0, 0,)
        if episode % save_freq == 0:
            agent.save_model(
                f"sac_actor_episode{episode}_\
                {env_id}.h5",
                f"sac_critic_episode{episode}_\
                {env_id}.h5",
            )
```

9. Now for the crucial steps! Let's use the agent's `act` and `train` methods to collect experience by acting (take actions) and training the agent using the collected experience:

```
        if epoch > random_epochs:
            use_random = False

        action = agent.act(np.array(cur_state),
                        use_random=use_random)
        next_state, reward, done, _ = client.env_step(
            instance_id, action.numpy().tolist()
        )
        agent.train(np.array(cur_state), action, reward,
                    np.array(next_state), done)
```

10. Let's now update the variables to prepare for the next step in the episode:

```
        cur_state = next_state
        episode_reward += reward
        steps += 1
        epoch += 1


        # Update Tensorboard with Agent's training status
        agent.log_status(summary_writer, epoch, reward)
        summary_writer.flush()
```

11. That completes our training loop. It was simple, wasn't it? Let's not forget to save the agent's model after training so that we can use the trained model when it's time for deployment:

```
agent.save_model(
    f"sac_actor_final_episode_{env_id}.h5", \
    f"sac_critic_final_episode_{env_id}.h5"
)
```

12. You can now proceed and run the script using the following command:

```
(tfrl-cookbook)praveen@desktop:~/tensorflow2-
reinforcement-learning-cookbook/src/ch7-cloud-deploy-
deep-rl-agents$ python 3_training_rl_agents_using_remote_
sims.py
```

13. Did we forget something? Which sim server is the client connecting to? Is the sim server running?! If you get a long error on the command line that ends with a line that looks like the following, it certainly means that the sim server is not running:

```
Failed to establish a new connection: [Errno 111]
Connection refused'))
```

14. Let's do it correctly this time! Let's make sure our sim server is running by launching the tradegym server using the following command:

```
(tfrl-cookbook)praveen@desktop:~/tensorflow2-
reinforcement-learning-cookbook/src/ch7-cloud-deploy-
deep-rl-agents$ python tradegym_http_server.py
```

15. We can now launch the agent training script using the following command (same as before):

```
(tfrl-cookbook)praveen@desktop:~/tensorflow2-
reinforcement-learning-cookbook/src/ch7-cloud-deploy-
deep-rl-agents$ python 3_training_rl_agents_using_remote_
sims.py
```

16. You should see an output similar to the following:

```
. . .
Total params: 16,257
Trainable params: 16,257
Non-trainable params: 0
```

```
None
episode:1 cumulative_reward:370.45421418744525 steps:9
epochs:9
episode:2 cumulative_reward:334.52956448599605 steps:9
epochs:18
episode:3 cumulative_reward:375.27432450733943 steps:9
epochs:27
episode:4 cumulative_reward:363.7160827166332 steps:9
epochs:36
episode:5 cumulative_reward:363.2819222532322 steps:9
epochs:45
. . .
```

That completes our script for training RL agents using remote sims!

How it works...

So far, we have been directly using the gym library to interact with the sim since we were running the RL environment simulators as part of the agent training scripts. While this will be good enough for CPU-bound local simulators, as we start to use advanced simulators or simulators that we don't own, or even those cases where we don't want to run or manage the simulator instances, we can leverage the client wrapper we built using our previous recipe in this chapter and talk to tradegym-like RL environments that expose a REST API for interfacing. In this recipe, the agent training script utilizes the tradegym client module to interact with a remote tradegym server to complete the RL training loop.

With that, let's move on to the next recipe to see how we can evaluate previously trained agents.

Testing/evaluating RL agents

Let's assume that you have trained the SAC agent in one of the trading environments using the training script (previous recipe) and that you have several versions of the trained agent models, each with different policy network architectures or hyperparameters or your own tweaks and customizations to improve its performance. When you want to deploy an agent, you want to make sure that you pick the best performing agent, don't you?

This recipe will help you build a lean script to evaluate a given pre-trained agent model locally so that you can get a quantitative performance assessment and compare several trained models before choosing the right agent model for deployment. Specifically, we will use the `tradegym` module and the `sac_agent_runtime` module that we built earlier in this chapter to evaluate the agent models that we train.

Let's get started!

Getting ready

To complete this recipe, you will first need to activate the `tf2rl-cookbook` Python/ conda virtual environment. Make sure to update the environment to match the latest conda environment specification file (`tfrl-cookbook.yml`) in the cookbook's code repository. If the following `import` statements run without issues, you are ready to get started:

```
#!/bin/env/python
import os
import sys


from argparse import ArgumentParser
import imageio
import gym
```

How to do it...

Let's focus on creating a simple, but complete, agent evaluation script:

1. First, let's import the `tradegym` module for the training environment and the SAC agent runtime:

   ```
   sys.path.append(os.path.dirname(os.path.abspath(__
   file__)))
   import tradegym  # Register tradegym envs with OpenAI Gym
   registry
   from sac_agent_runtime import SAC
   ```

2. Next, let's create a command-line argument parser to handle command-line configurations:

```
parser = ArgumentParser(prog="TFRL-Cookbook-Ch7-
Evaluating-RL-Agents")
parser.add_argument("--agent", default="SAC", help="Name
of Agent. Default=SAC")
```

3. Let's now add support for the --env argument to specify the RL environment ID and --num-episodes to specify the number of episodes for evaluating the agent. Let's have some sensible default values for both the arguments so that we can run the script even without any arguments for quick (or lazy?!) testing:

```
parser.add_argument(
    "--env",
    default="StockTradingContinuousEnv-v0",
    help="Name of Gym env.
Default=StockTradingContinuousEnv-v0",
)
parser.add_argument(
    "--num-episodes",
    default=10,
    help="Number of episodes to evaluate the agent.\
        Default=100",
)
```

4. Let's also add support for --trained-models-dir to specify the directory containing the trained models, and the --model-version flag to specify the specific version of the model in that trained directory:

```
parser.add_argument(
    "--trained-models-dir",
    default="trained_models",
    help="Directory contained trained models.
Default=trained_models",
)
parser.add_argument(
    "--model-version",
    default="episode100",
```

```
            help="Trained model version. Default=episode100",
    )
```

5. We are now ready to finalize the argument parsing:

```
args = parser.parse_args()
```

6. Let's begin our implementation of the __main__ method in this step and continue its implementation in the following steps. Let's start by creating a local instance of the RL environment in which we want to evaluate the agent:

```
if __name__ == "__main__":
    # Create an instance of the evaluation environment
    env = gym.make(args.env)
```

7. Let's now initialize the agent class. For now, we only support the SAC agent, but it is quite easy to add support for other agents we have discussed in this book:

```
if args.agent != "SAC":
    print(f"Unsupported Agent: {args.agent}. Using \
        SAC Agent")
    args.agent = "SAC"
# Create an instance of the Soft Actor-Critic Agent
agent = SAC(env.observation_space.shape, \
        env.action_space)
```

8. Next, let's load the trained agent models:

```
# Load trained Agent model/brain
model_version = args.model_version
agent.load_actor(
    os.path.join(args.trained_models_dir, \
            f"sac_actor_{model_version}.h5")
)
agent.load_critic(
    os.path.join(args.trained_models_dir, \
            f"sac_critic_{model_version}.h5")
)
print(f"Loaded {args.agent} agent with trained \
        model version:{model_version}")
```

9. We are now ready to evaluate the agent using the trained models in the test environment:

```
# Evaluate/Test/Rollout Agent with trained model/
# brain
video = imageio.get_writer("agent_eval_video.mp4",\
                            fps=30)
avg_reward = 0
for i in range(args.num_episodes):
    cur_state, done, rewards = env.reset(), False, 0
    while not done:
        action = agent.act(cur_state, test=True)
        next_state, reward, done, _ = \
                            env.step(action[0])
        cur_state = next_state
        rewards += reward
        if render:
            video.append_data(env.render(mode=\
                            "rgb_array"))
    print(f"Episode#:{i} cumulative_reward:\
            {rewards}")
    avg_reward += rewards
avg_reward /= args.num_episodes
video.close()
print(f"Average rewards over {args.num_episodes} \
        episodes: {avg_reward}")
```

10. Let's now try to evaluate the agent in the StockTradingContinuous-v0 environment. Note that the market data source for the stocks trading environment in data/MSFT.csv and data/TSLA.csv can be different from the market data used for training! After all, we want to evaluate how well the agent has learned to trade! Run the following command to launch the agent evaluation script:

```
(tfrl-cookbook)praveen@desktop:~/tensorflow2-
reinforcement-learning-cookbook/src/ch7-cloud-deploy-
deep-rl-agents$ python 4_evaluating_rl_agents.py
```

11. Depending on how well your trained agents are, you will see output on the console similar to the following (reward values will vary):

```
. . .

========================================================
=========================================
Total params: 16,257
Trainable params: 16,257
Non-trainable params: 0


_____
_____

None
Loaded SAC agent with trained model version:episode100
Episode#:0 cumulative_reward:382.5117154452246
Episode#:1 cumulative_reward:359.27720004181674
Episode#:2 cumulative_reward:370.92829808499664
Episode#:3 cumulative_reward:341.44002189086007
Episode#:4 cumulative_reward:364.32631211784394
Episode#:5 cumulative_reward:385.89219327764476
Episode#:6 cumulative_reward:365.2120387185878
Episode#:7 cumulative_reward:339.98494537310785
Episode#:8 cumulative_reward:362.7133769241483
Episode#:9 cumulative_reward:379.12388043270073
Average rewards over 10 episodes:  365.1409982306931
. . .
```

That's it!

How it works...

We initialized an SAC agent with only the runtime components that are required to evaluate the agent using the sac_agent_runtime module and then loaded previously trained model versions (for both the actor and the critic), which are both customizable using the command-line arguments. We then created a local instance of the StockTradingContinuousEnv-v0 environment using the tradegym library and evaluated our agents to get the cumulative reward as one of the quantitative indicators of the performance of the trained agent models.

Now that we know how to evaluate and pick the best performing agent, let's move on to the next recipe to understand how to package the trained agent for deployment!

Packaging RL agents for deployment – a trading bot

This is one of the crucial recipes of this chapter, where we will be discussing how to package the agent so that we can deploy on the cloud (next recipe!) as a service. We will be implementing a script that takes our trained agent models and exposes the `act` method as a RESTful service. We will then package the agent and the API script into a **Docker** container that is ready to be deployed to the cloud! By the end of this recipe, you will have built a deployment-ready Docker container with your trained RL agent that is ready to create and offer your Agent/Bot-as-a-Service!

Let's jump into the details.

Getting ready

To complete this recipe, you will first need to activate the `tf2rl-cookbook` Python/ conda virtual environment. Make sure to update the environment to match the latest conda environment specification file (`tfrl-cookbook.yml`) in the cookbook's code repository. If the following `import` statements run without issues, you are ready to undertake the next step, which is to set up Docker:

```
import os
import sys
from argparse import ArgumentParser

import gym.spaces
from flask import Flask, request
import numpy as np
```

For this recipe, you will need Docker installed. Please follow the official setup instructions to install Docker for your platform. You can find the instructions at `https://docs. docker.com/get-docker/`.

How to do it...

We will first implement the script to expose the agent's `act` method as a REST service and then move on to create the Dockerfile for containerizing the agent:

1. First, let's import the `sac_agent_runtime` that we built earlier in this chapter:

```
sys.path.append(os.path.dirname(os.path.abspath(__
file__)))
from sac_agent_runtime import SAC
```

2. Next, let's create a handler for the command-line argument and `--agent` as the first supported argument to allow specification of the agent algorithm we want to use:

```
parser = ArgumentParser(
    prog="TFRL-Cookbook-Ch7-Packaging-RL-Agents-For-
Cloud-Deployments"
)
parser.add_argument("--agent", default="SAC", help="Name
of Agent. Default=SAC")
```

3. Next, let's add arguments to allow specification of the IP address and the port of the host server where our agent will be deployed. We will set and use the defaults for now and can change them from the command line when we need to:

```
parser.add_argument(
    "--host-ip",
    default="0.0.0.0",
    help="IP Address of the host server where Agent
        service is run. Default=127.0.0.1",
)
parser.add_argument(
    "--host-port",
    default="5555",
    help="Port on the host server to use for Agent
        service. Default=5555",
)
```

4. Next, let's add support for arguments to specify the directory containing the trained agent models and the specific model version to use:

```
parser.add_argument(
    "--trained-models-dir",
    default="trained_models",
    help="Directory contained trained models. \
        Default=trained_models",
)
parser.add_argument(
    "--model-version",
    default="episode100",
    help="Trained model version. Default=episode100",
)
```

5. As the final set of supported arguments, let's add arguments to allow specification of the observation shape and the action space specifications based on the trained model configuration:

```
parser.add_argument(
    "--observation-shape",
    default=(6, 31),
    help="Shape of observations. Default=(6, 31)",
)
parser.add_argument(
    "--action-space-low", default=[-1], help="Low value \
      of action space. Default=[-1]"
)
parser.add_argument(
    "--action-space-high", default=[1], help="High value\
      of action space. Default=[1]"
)
parser.add_argument(
    "--action-shape", default=(1,), help="Shape of \
    actions. Default=(1,)"
)
```

6. We can now finalize the argument parser and start the implementation of the __main__ function:

```
args = parser.parse_args()


if __name__ == "__main__":
```

7. First, let's load the runtime configurations for the agent:

```
if args.agent != "SAC":
    print(f"Unsupported Agent: {args.agent}. Using \
        SAC Agent")
    args.agent = "SAC"
# Set Agent's runtime configs
observation_shape = args.observation_shape
action_space = gym.spaces.Box(
    np.array(args.action_space_low),
    np.array(args.action_space_high),
    args.action_shape,
)
```

8. Next, let's create an instance of the agent and load the weights for the agent's actor and critic networks from the pre-trained model:

```
# Create an instance of the Agent
agent = SAC(observation_shape, action_space)
# Load trained Agent model/brain
model_version = args.model_version
agent.load_actor(
    os.path.join(args.trained_models_dir, \
        f"sac_actor_{model_version}.h5")
)
agent.load_critic(
    os.path.join(args.trained_models_dir, \
        f"sac_critic_{model_version}.h5")
)
print(f"Loaded {args.agent} agent with trained model\
    version:{model_version}")
```

9. We can now set up the service endpoint using Flask, which is going to be as simple as shown in the following lines of code. Note that we are exposing the agent's act method at the /v1/act endpoint:

```
# Setup Agent (http) service
app = Flask(__name__)
@app.route("/v1/act", methods=["POST"])
def get_action():
    data = request.get_json()
    action = agent.act(np.array(data.get(
                        "observation")), test=True)
    return {"action": action.numpy().tolist()}
```

10. Finally, we just have to add the line that launches the Flask application to start the service when executed:

```
# Launch/Run the Agent (http) service
app.run(host=args.host_ip, port=args.host_port,
        debug=True)
```

11. Our REST API implementation for the agent is ready. We can now focus on creating a Docker container for the agent service. We will start to implement the Dockerfile by specifying the base image to be nvidia/cuda:* so that we have the GPU drivers that are necessary to utilize the GPU on the server where we will be deploying the agent. The following code lines in this and the following steps will go into a file named Dockerfile:

```
FROM nvidia/cuda:10.1-cudnn7-devel-ubuntu18.04
# TensorFlow2.x Reinforcement Learning Cookbook
# Chapter 7: Deploying Deep RL Agents to the cloud
LABEL maintainer="emailid@domain.tld"
```

12. Let's now install a few system-level packages that are necessary and clean up the files to save disk space:

```
RUN apt-get install -y wget git make cmake zlib1g-dev &&
rm -rf /var/lib/apt/lists/*
```

13. To execute our agent runtime with all the necessary packages installed, we will make use of the conda Python environment. So, let's go ahead and go through the instructions to download and set up `miniconda` in the container:

```
ENV PATH="/root/miniconda3/bin:${PATH}"
ARG PATH="/root/miniconda3/bin:${PATH}"
RUN apt-get update
RUN wget \
    https://repo.anaconda.com/miniconda/Miniconda3-
latest-Linux-x86_64.sh \
    && mkdir /root/.conda \
    && bash Miniconda3-latest-Linux-x86_64.sh -b \
    && rm -f Miniconda3-latest-Linux-x86_64.sh
# conda>4.9.0 is required for `--no-capture-output`
RUN conda update -n base conda
```

14. Let's now copy the source code of this chapter into the container and create the conda environment using the list of packages specified in the `tfrl-cookbook.yml` file:

```
ADD . /root/tf-rl-cookbook/ch7
WORKDIR /root/tf-rl-cookbook/ch7
RUN conda env create -f "tfrl-cookbook.yml" -n "tfrl-
cookbook"
```

15. Finally, we just have to set the `ENTRYPOINT` for the container and the `CMD`, which will be passed as arguments to the `ENTRYPOINT` when the container is launched:

```
ENTRYPOINT [ "conda", "run", "--no-capture-output", "-n",
"tfrl-cookbook", "python" ]
CMD [ "5_packaging_rl_agents_for_deployment.py" ]
```

16. That completes our Dockerfile and we are now ready to package our agent by building the Docker container. You can run the following command to build the Docker container as per the instructions in the Dockerfile and tag it with a container image name of your choice. Let's use the following command:

```
(tfrl-cookbook)praveen@desktop:~/tensorflow2-
reinforcement-learning-cookbook/src/ch7-cloud-deploy-
deep-rl-agents$docker build -f Dockerfile -t tfrl-
cookbook/ch7-trading-bot:latest
```

17. If you are running the preceding command for the first time, it may take quite some
    time for Docker to build the container. Subsequent runs or updates will run faster
    as intermediate layers might already have been cached when run for the first time.
    When things are running smoothly, you will see an output similar to the following
    (note that most of the layers are cached since I had already built the container
    previously):

```
Sending build context to Docker daemon  1.793MB
Step 1/13 : FROM nvidia/cuda:10.1-cudnn7-devel-
ubuntu18.04
 ---> a3bd8cb789b0
Step 2/13 : LABEL maintainer="emailid@domain.tld"
 ---> Using cache
 ---> 4322623c24c8
Step 3/13 : ENV PATH="/root/miniconda3/bin:${PATH}"
 ---> Using cache
 ---> e9e8c882662a
Step 4/13 : ARG PATH="/root/miniconda3/bin:${PATH}"
 ---> Using cache
 ---> 31d45d5bcb05
Step 5/13 : RUN apt-get update
 ---> Using cache
 ---> 3f7ed3eb3c76
Step 6/13 : RUN apt-get install -y wget git make cmake
zlib1g-dev && rm -rf /var/lib/apt/lists/*
 ---> Using cache
 ---> 0ffb6752f5f6
Step 7/13 : RUN wget       https://repo.anaconda.com/
miniconda/Miniconda3-latest-Linux-x86_64.sh       && mkdir
/root/.conda      && bash Miniconda3-latest-Linux-x86_64.
sh -b       && rm -f Miniconda3-latest-Linux-x86_64.sh
 ---> Using cache
```

18. For the layers that involve COPY/ADD file operations from the disk, the instructions will be run as they cannot be cached. For example, you will see that the following steps from *step 9* will continue to be run fresh without using any cache. This is normal even if you have already built the container:

```
Step 9/13 : ADD . /root/tf-rl-cookbook/ch7
 ---> ed8541c42ebc
Step 10/13 : WORKDIR /root/tf-rl-cookbook/ch7
 ---> Running in f5a9c6ad485c
Removing intermediate container f5a9c6ad485c
 ---> 695ca00c6db3
Step 11/13 : RUN conda env create -f "tfrl-cookbook.yml"
-n "tfrl-cookbook"
 ---> Running in b2a9706721e7
Collecting package metadata (repodata.json):
...working... done
Solving environment: ...working... done...
```

19. Finally, when the Docker container is built, you will see something like the following message:

```
Step 13/13 : CMD [ "2_packaging_rl_agents_for_deployment.
py" ]
 ---> Running in 336e442b0218
Removing intermediate container 336e442b0218
 ---> cc1caea406e6
Successfully built cc1caea406e6
Successfully tagged tfrl-cookbook/ch7:latest
```

Congratulations on successfully packing your RL agent for deployment!

How it works...

We leveraged the `sac_agent_runtime` we built earlier in this chapter to create and initialize an SAC agent instance. We then loaded pre-trained agent models for both the actor and the critic. After that, we exposed the SAC agent's `act` method as a REST API with an HTTP POST endpoint for taking the observations as the POST message and returning actions as the response to the POST request. Finally, we launched the script as a Flask application to start serving.

In the second part of the recipe, we packaged the agent actioa-serving application as a Docker container and prepared it for deployment!

We are now on the verge of deploying the agent to the cloud! March on to the next recipe and find out how.

Deploying RL agents to the cloud – a trading Bot-as-a-Service

The ultimate goal of training an RL agent is to use it for taking actions given new observations. In the case of our stock trading SAC agent, we have so far learned to train, evaluate, and package the best performing agent model to build our trading bot. While we focused on one particular application (autonomous trading bot), you can see how easy it is to change the training environment or agent algorithms based on the recipes in earlier chapters of this book. This recipe will walk you through the steps to deploy the Docker containerized/packaged RL agent to the cloud and run the Bot-as-a-Service.

Getting ready

To complete this recipe, you will need access to a cloud service such as Azure, AWS, GCP, Heroku or another cloud service provider that allows you to host and run your Docker container. If you are a student, you can make use of GitHub's student developer pack (`https://education.github.com/pack`), which, as of 2020, allows you to avail yourself of several benefits for free, including $100 worth of Microsoft Azure credits or $50 platform credit on DigitalOcean if you are a new user.

Several guides exist on how to push your Docker container to the cloud and to deploy/run the container as a service. For example, if you have an Azure account, you can following the official guide here: `https://docs.microsoft.com/en-us/azure/container-instances/container-instances-quickstart`.

The guide walks you through the various options (the CLI, Portal, PowerShell, ARM templates, and the Docker CLI) to deploy your Docker container-based agent service.

How to do it...

We will first deploy the trading bot locally and test it out. After that, we can deploy it to your cloud service of choice. As an example, this recipe will walk you through the steps to deploy it to Heroku (`https://heroku.com`).

Let's begin:

1. Let's first build our Docker container containing our trading bot using the following command. Note that if you have already built the container by following the previous recipe in this chapter, the following command may finish executing faster depending on the cached layers and your changes to the Dockerfile:

    ```
    (tfrl-cookbook)praveen@desktop:~/tensorflow2-
    reinforcement-learning-cookbook/src/ch7-cloud-deploy-
    deep-rl-agents$docker build -f Dockerfile -t tfrl-
    cookbook/ch7-trading-bot:latest
    ```

2. Once the Docker container with the bot has been built successfully, we can launch the bot using the following command:

    ```
    (tfrl-cookbook)praveen@desktop:~/tensorflow2-
    reinforcement-learning-cookbook/src/ch7-cloud-deploy-
    deep-rl-agents$docker run -it -p 5555:5555 tfrl-cookbook/
    ch7-trading-bot
    ```

3. If all goes well, you should see console output somewhat similar to the following, indicating that the bot is up and running and ready to act:

    ```
    . . .
    =================================================================
    ==========================================
    Total params: 16,257
    Trainable params: 16,257
    Non-trainable params: 0


    None
    Loaded SAC agent with trained model version:episode100
     * Debugger is active!
     * Debugger PIN: 604-104-903

    . . .
    ```

4. Now that you have deployed the trading bot locally (on your own server), let's create a simple script to utilize the Bot-as-a-Service that you have built. Create a file named `test_agent_service.py` with the following content:

```
#Simple test script for the deployed Trading Bot-as-a-
Service
import os
import sys
import gym
import requests
sys.path.append(os.path.dirname(os.path.abspath(__
file__)))
import tradegym  # Register tradegym envs with OpenAI Gym
# registry
host_ip = "127.0.0.1"
host_port = 5555
endpoint = "v1/act"
env = gym.make("StockTradingContinuousEnv-v0")
post_data = {"observation": env.observation_space.
sample().tolist()}
res = requests.post(f"http://{host_ip}:{host_port}/
{endpoint}", json=post_data)
if res.ok:
    print(f"Received Agent action:{res.json()}")
```

5. You can execute the script using the following command:

```
(tfrl-cookbook)praveen@desktop:~/tensorflow2-
reinforcement-learning-cookbook/src/ch7-cloud-deploy-
deep-rl-agents$python test_agent_service.py
```

6. Note that your bot container still needs to be running. Once you execute the preceding command, you will see an output similar to the following line on the console output of your bot, indicating that a new POST message was received at the `/v1/act` endpoint, which was served with an HTTP response status of 200, indicating success:

```
172.17.0.1 - - [00/Mmm/YYYY hh:mm:ss] "POST /v1/act
HTTP/1.1" 200 -
```

7. You will also notice that your test script will print out an output similar to the following on its console window, indicating that it received an action from the trading bot:

```
Received Agent action:{'action': [[0.008385116065491426]]}
```

8. It's time to deploy your trading bot to a cloud platform so that you or others can access it over the internet! As discussed in the *Getting started* section, you have several options in terms of choosing the cloud provider where you want to host your Docker container image and deploy the RL agent Bot-as-a-Service. We will use Heroku as an example as it offers free hosting and a simple CLI. First, you need to install the Heroku CLI. Follow the official instructions listed at `https://devcenter.heroku.com/articles/heroku-cli` to install the Heroku CLI for your platform (Linux/Windows/macOS X). On Ubuntu Linux, we can use the following command:

```
sudo snap install --classic heroku
```

9. Once the Heroku CLI is installed, you can log in to the Heroku container registry using the following command:

```
heroku container:login
```

10. Next, run the following command from the directory containing the Dockerfile for the agent; for example:

```
tfrl-cookbook)praveen@desktop:~/tensorflow2-
reinforcement-learning-cookbook/src/ch7-cloud-deploy-
deep-rl-agents$heroku create
```

11. If you have not logged in to Heroku, you will be prompted to log in:

```
Creating app... !
        Invalid credentials provided.
    >   Warning: heroku update available from 7.46.2 to
7.47.0.
heroku: Press any key to open up the browser to login or
q to exit:
```

12. Once you are logged in, you will get an output similar to the following:

```
Creating salty-fortress-4191... done, stack is heroku-18
https://salty-fortress-4191.herokuapp.com/ | https://git.
heroku.com/salty-fortress-4191.git
```

13. That's the address of your container registry on Heroku. You can now build your bot container and push it to Heroku using the following command:

```
heroku container:push web
```

14. Once that process completes, you can release the bot container image to your Heroku app using the following command:

```
heroku container:release web
```

15. Congratulations! You have just deployed your bot to the cloud! You can now access your bot at the new address, such as https://salty-fortress-4191. herokuapp.com/ in the sample used in the preceding code. You should be able to send observations to your bot and get the actions in return! Congratulations on deploying your Bot-as-a-Service!

We are now ready to wrap up the chapter.

How it works...

We first built and launched the Docker container locally on your machine by using the docker run command and specifying that you want the local port 5555 to be mapped with the container's port 5555. This will allow the host (your machine) to communicate with your container using that port as if it were a local port on the machine. Following deployment, we used a test script that uses the Python request library to create a POST request with sample data for the observation values and sent it to the bot running in the container. We observed how the bot responded to the request via the command-line status output followed by a returned success response containing the bot's trading action.

We then deployed the same container with the bot to the cloud (Heroku). Following a successful deployment, the bot was accessible over the web using the public herokuapp URL created automatically by Heroku.

That completes the recipe and the chapter! I hope you enjoyed working through it. See you in the next chapter.

8

Distributed Training for Accelerated Development of Deep RL Agents

Training Deep RL agents to solve a task takes enormous wall-clock time due to the high sample complexity. For real-world applications, iterating over agent training and testing cycles at a faster pace plays a crucial role in the market readiness of a Deep RL application. The recipes in this chapter provide instructions on how to speed up Deep RL agent development using the distributed training of deep neural network models by leveraging TensorFlow 2.x's capabilities. Strategies for utilizing multiple CPUs and GPUs both on a single machine and across a cluster of machines are discussed. Multiple recipes for training distributed **Deep Reinforcement Learning (Deep RL)** agents using the **Ray, Tune**, and **RLLib** frameworks are also provided.

Specifically, the following recipes are a part of this chapter:

- Building distributed deep learning models using TensorFlow 2.x – Multi-GPU training
- Scaling up and out – Multi-machine, multi-GPU training
- Training Deep RL agents at scale – Multi-GPU PPO agent
- Building blocks for distributed Deep Reinforcement Learning for accelerated training
- Large-scale Deep RL agent training using Ray, Tune, and RLLib

Technical requirements

The code in the book is extensively tested on Ubuntu 18.04 and Ubuntu 20.04 and should work with later versions of Ubuntu if Python 3.6+ is available. With Python 3.6+ installed along with the necessary Python packages, as listed before the start of each of the recipes, the code should run fine on Windows and Mac OSX too. It is advised to create and use a Python virtual environment named `tf2rl-cookbook` to install the packages and run the code in this book. Miniconda or Anaconda installation for Python virtual environment management is recommended.

The complete code for each recipe in each chapter will be available here: `https://github.com/PacktPublishing/Tensorflow-2-Reinforcement-Learning-Cookbook`.

Distributed deep learning models using TensorFlow 2.x – Multi-GPU training

Deep RL utilizes a deep neural network for policy, value-function, or model representations. For higher-dimensional observation/state spaces, for example, in the case of image or image-like observations, it is typical to use **convolutional neural network (CNN)** architectures. While CNNs are powerful and enable training Deep RL policies for vision-based control tasks, training deep CNNs requires a lot of time, especially in the RL setting. This recipe will help you understand how we can leverage TensorFlow 2.x's distributed training APIs to train deep **residual networks (ResNets)** using multiple GPUs. The recipe comes with configurable building blocks that you can use to build Deep RL components like deep policy networks or value networks.

Let's get started!

Getting ready

To complete this recipe, you will first need to activate the tf2rl-cookbook Python/ conda virtual environment. Make sure to update the environment to match the latest conda environment specification file (tfrl-cookbook.yml) in the cookbook's code repo. Having access to a (local or cloud) machine with one or more GPUs will be beneficial for this recipe. We will be using the **TensorFlow Datasets** library in this recipe, which is available as a Python package named tensorflow_datasets. This should be already installed if you used tfrl-cookbook.yml to set up/update your conda environment.

Now, let's begin!

How to do it...

The implementation in this recipe is based on the latest official TensorFlow documentation/tutorial. The following steps will help you get a good command over TensorFlow 2.x's distributed execution capabilities. We will be using a ResNet model as an example of a large model that will benefit from being trained in a distributed fashion, utilizing multiple GPUs to speed up training. We will discuss the code snippets for the main components for building a ResNet. Please refer to the resnet.py file in the cookbook's code repository for the full and complete implementation. Let's get started:

1. Let's jump right into the template for building residual neural networks:

```python
def resnet_block(
    input_tensor, size, kernel_size, filters, stage, \
     conv_strides=(2, 2), training=None
):
    x = conv_building_block(
        input_tensor,
        kernel_size,
        filters,
        stage=stage,
        strides=conv_strides,
        block="block_0",
        training=training,
    )
    for i in range(size - 1):
        x = identity_building_block(
            x,
```

```
                kernel_size,
                filters,
                stage=stage,
                block="block_%d" % (i + 1),
                training=training,
            )
        return x
```

2. With the above template for a ResNet block, we can quickly build ResNets with multiple ResNet blocks. We will implement a ResNet with one ResNet block here in the book, and you will find the ResNet implemented with multiple configurable numbers and sizes of ResNet blocks in the code repository. Let's get started and complete the ResNet implementation in the following several steps, focusing on one important concept at a time. First, let's define the function signature:

```
def resnet(num_blocks, img_input=None, classes=10,
training=None):
    """Builds the ResNet architecture using provided
        config"""
```

3. Next, let's handle the channel ordering in the input image data representation. The most common ordering of the dimensions is either: batch_size x channels x width x height or batch_size x width x height x channels. We will handle these two cases:

```
if backend.image_data_format() == "channels_first":
    x = layers.Lambda(
        lambda x: backend.permute_dimensions(x, \
            (0, 3, 1, 2)), name="transpose"
    )(img_input)
    bn_axis = 1
else:  # channel_last
    x = img_input
    bn_axis = 3
```

4. Now, let's apply zero padding to the input and apply initial layers to start processing:

```
x = tf.keras.layers.ZeroPadding2D(padding=(1, 1), \
                                name="conv1_pad")(x)
x = tf.keras.layers.Conv2D(16,(3, 3),strides=(1, 1),
                    padding="valid",
                    kernel_initializer="he_normal",
                    kernel_regularizer= \
                        tf.keras.regularizers.l2(
                            L2_WEIGHT_DECAY),
                    bias_regularizer= \
                        tf.keras.regularizers.l2(
                            L2_WEIGHT_DECAY),
                    name="conv1",)(x)
x = tf.keras.layers.BatchNormalization(axis=bn_axis,
            name="bn_conv1", momentum=BATCH_NORM_DECAY,
            epsilon=BATCH_NORM_EPSILON,)\
                (x, training=training)
x = tf.keras.layers.Activation("relu")(x)
```

5. It's time to add the ResNet blocks using the resnet_block function we created:

```
x = resnet_block(x, size=num_blocks, kernel_size=3,
        filters=[16, 16], stage=2, conv_strides=(1, 1),
        training=training,)
x = resnet_block(x, size=num_blocks, kernel_size=3,
        filters=[32, 32], stage=3, conv_strides=(2, 2),
        training=training)
x = resnet_block(x, size=num_blocks, kernel_size=3,
        filters=[64, 64], stage=4, conv_strides=(2, 2),
        training=training,)
```

6. As the final layer, we want to add a `softmax` activated `Dense` (fully connected) layer with the number of nodes equal to the number of output classes needed for the task:

```
x = tf.keras.layers.GlobalAveragePooling2D(
                                 name="avg_pool")(x)
    x = tf.keras.layers.Dense(classes,
        activation="softmax",
        kernel_initializer="he_normal",
        kernel_regularizer=tf.keras.regularizers.l2(
            L2_WEIGHT_DECAY),
        bias_regularizer=tf.keras.regularizers.l2(
            L2_WEIGHT_DECAY),
        name="fc10",)(x)
```

7. The last step in the ResNet model building function is to wrap the layers as a TensorFlow 2.x Keras model and return the output:

```
    inputs = img_input
    # Create model.
    model = tf.keras.models.Model(inputs, x,
  name=f"resnet{6 * num_blocks + 2}")


    return model
```

8. Using the ResNet function that we just discussed, it becomes quite easy to build deep residual networks of varying layer depths by simply changing the number of blocks. For example, the following is possible:

```
resnet_mini = functools.partial(resnet, num_blocks=1)
resnet20 = functools.partial(resnet, num_blocks=3)
resnet32 = functools.partial(resnet, num_blocks=5)
resnet44 = functools.partial(resnet, num_blocks=7)
resnet56 = functools.partial(resnet, num_blocks=9)
```

9. With our model defined, we can jump to the multi-GPU training code. The remaining steps in this recipe will guide you through the implementation that will allow you to speed up training the ResNet using all the available GPUs on a machine. Let's start by importing the `ResNet` module that we built along with the `tensorflow_datasets` module:

```
import os
import sys
import tensorflow as tf
import tensorflow_datasets as tfds


if "." not in sys.path:
    sys.path.insert(0, ".")
import resnet
```

10. We can now choose which dataset we want to use to exercise our distributed training pipeline. For this recipe, we will use the `dmlab` dataset that contains images typically observed by RL agents acting in the DeepMind Lab environment. Depending on the compute capabilities of the GPUs, RAM, and CPUs on your training machine, you may want to use a smaller dataset such as `CIFAR10`:

```
dataset_name = "dmlab"  # "cifar10" or "cifar100"; See
tensorflow.org/datasets/catalog for complete list
# NOTE: dmlab is large in size; Download bandwidth and #
GPU memory to be considered
datasets, info = tfds.load(name="dmlab", with_info=True,
                           as_supervised=True)
dataset_train, dataset_test = datasets["train"], \
                           datasets["test"]
input_shape = info.features["image"].shape
num_classes = info.features["label"].num_classes
```

11. The next step needs your full attention! We are going to choose the distributed execution strategy. TensorFlow 2.x has wrapped a lot of functionality into a simple API call like the one listed here:

```
strategy = tf.distribute.MirroredStrategy()
print(f"Number of devices: {
        strategy.num_replicas_in_sync}")
```

12. We will declare the key hyperparameters in this step that you can adjust depending on your machine's hardware (such as RAM and GPU memory):

```
num_train_examples = info.splits["train"].num_examples
num_test_examples = info.splits["test"].num_examples

BUFFER_SIZE = 1000  # Increase as per available memory
BATCH_SIZE_PER_REPLICA = 64
BATCH_SIZE = BATCH_SIZE_PER_REPLICA * \
                strategy.num_replicas_in_sync
```

13. Before we start preparing the datasets, let's implement a preprocessing function that performs operations before we pass the images to the neural network. You can add your own custom preprocessing operations. In this recipe, we will only need to cast the image data to float32 first and then convert the image pixel value ranges to be [0, 1] rather than the typical interval of [0, 255]:

```
def preprocess(image, label):
    image = tf.cast(image, tf.float32)
    image /= 255
    return image, label
```

14. We are ready to create the dataset splits for training and validation/testing:

```
train_dataset = (
    dataset_train.map(preprocess).cache().\
        shuffle(BUFFER_SIZE).batch(BATCH_SIZE)
)
eval_dataset = dataset_test.map(preprocess).batch(
                                                BATCH_SIZE)
```

15. We are at the crucial step of this recipe! Let's instantiate and compile our model within the scope of the distributed strategy:

```
with strategy.scope():
    # model = create_model()
    model = create_model("resnet_mini")
    tf.keras.utils.plot_model(model,
```

```
                             to_file="./slim_resnet.png",
                             show_shapes=True)
    model.compile(
        loss=\
            tf.keras.losses.SparseCategoricalCrossentropy(
                from_logits=True),
        optimizer=tf.keras.optimizers.Adam(),
        metrics=["accuracy"],
    )
```

16. Let's also create callbacks for logging to TensorBoard and checkpointing our model parameters during training:

```
checkpoint_dir = "./training_checkpoints"
checkpoint_prefix = os.path.join(checkpoint_dir,
                                "ckpt_{epoch}")
callbacks = [
    tf.keras.callbacks.TensorBoard(
        log_dir="./logs", write_images=True, \
        update_freq="batch"
    ),
    tf.keras.callbacks.ModelCheckpoint(
        filepath=checkpoint_prefix, \
        save_weights_only=True
    ),
]
```

17. With that, we have everything needed to train our model using the distributed strategy. With Keras's user-friendly fit() API, it is as simple as the following:

```
model.fit(train_dataset, epochs=12, callbacks=callbacks)
```

18. When the preceding line is executed, the training process will start. We can also manually save the model using the following lines:

```
path = "saved_model/"
model.save(path, save_format="tf")
```

19. Once we have a saved checkpoint, it is easy to load the weights and start evaluating the model:

```
model.load_weights(tf.train.latest_checkpoint(checkpoint_
dir))
eval_loss, eval_acc = model.evaluate(eval_dataset)
print("Eval loss: {}, Eval Accuracy: {}".format(eval_
loss, eval_acc))
```

20. To verify that the trained model using the distributed strategy works with and without replication, we will load it using two different methods in the following steps and evaluate. First, let's load the model without replicating using the (same) strategy we used to train the model:

```
unreplicated_model = tf.keras.models.load_model(path)


unreplicated_model.compile(
    loss=tf.keras.losses.\
        SparseCategoricalCrossentropy(from_logits=True),
    optimizer=tf.keras.optimizers.Adam(),
    metrics=["accuracy"],
)
eval_loss, eval_acc = unreplicated_model.evaluate(eval_
dataset)
print("Eval loss: {}, Eval Accuracy: {}".format(eval_
loss, eval_acc))
```

21. Next, let's load the model within the distributed execution strategy's scope, which would create replicas and evaluate the model:

```
with strategy.scope():
    replicated_model = tf.keras.models.load_model(path)
    replicated_model.compile(
        loss=tf.keras.losses.\
            SparseCategoricalCrossentropy(from_logits=True),
        optimizer=tf.keras.optimizers.Adam(),
        metrics=["accuracy"],

    )
```

```
eval_loss, eval_acc = \
    replicated_model.evaluate(eval_dataset)
print("Eval loss: {}, \
    Eval Accuracy: {}".format(eval_loss, eval_acc))
```

When you execute the preceding two code blocks, you will notice that both the methods result in the same evaluation accuracy, which is a good sign and signifies that we can use the model for prediction without any constraints on the execution strategy!

22. That completes our recipe. Let's recap and look at how the recipe works.

How it works...

A residual block in a neural network architecture applies convolution filters followed by multiple identity blocks. Specifically, a convolution block is applied once, followed by (size - 1) identity blocks where size is an integer representing the number of constituent convolutional-identity blocks. The identity block implements the shortcut or skip connections for the inputs to go through without being filtered through convolution operators. The convolutional block implements convolution layers followed by batch-normalization activation, followed by one or more sets of convolution-batchnorm-activation layers. The `resnet` module we built uses these convolution and identity building blocks to build a full ResNet with varying sizes that can be configured by simply changing the number of blocks. The size of the network is calculated as 6 * num_blocks + 2.

Once our ResNet model was ready, we used the `tensorflow_datasets` module to generate training and validation datasets. The TensorFlow Datasets module offers several popular datasets, such as CIFAR10, CIFAR100, and DMLAB, that have images and the associated labels for classification tasks. The list of all the available datasets can be found here: https://tensorflow.org/datasets/catalog.

In this recipe, we used the Mirrored Strategy for distributed execution using `tf.distribute.MirroredStrategy`, which enables synchronous distributed training using multiple replicas on one machine. Even with distributed execution with multiple replicas, we saw that the usual logging and checkpointing using callbacks worked as expected. We also verified that loading a saved model and running inference for evaluation works both with and without replication, making it portable without any added constraints just because the training utilized a distributed execution strategy!

It's time to advance to the next recipe!

Scaling up and out – Multi-machine, multi-GPU training

To reach the highest scale in terms of the distributed training of deep learning-based models, we need the capability to leverage compute resources across GPUs and across machines. This can significantly reduce the time it takes to iterate over or develop new models and architectures for the problem you are trying to solve. With easy access to cloud computing services such as Microsoft Azure, Amazon AWS, and Google's GCP, renting multiple GPU-equipped machines for an hourly rate has become easier and much more common. It is also more economical than setting up and maintaining your own multi-GPU multi-machine node. This recipe will provide a quick walk-through of training deep models using TensorFlow 2.x's multi-worker mirrored distributed execution strategy based on the official documentation, which you can use and easily customize for your use cases. For the multi-machine, multi-GPU distributed training example in this recipe, we will train a deep residual network (ResNet or resnet) for typical image classification tasks. The same network architecture can be used by RL agents for their policy or value-function representation with a slight modification to the output layer, as we will see in the later recipes of this chapter.

Let's get started!

Getting ready

To complete this recipe, you will first need to activate the `tf2rl-cookbook` Python/conda virtual environment. Make sure to update the environment to match the latest conda environment specification file (`tfrl-cookbook.yml`) in the cookbook's code repo. To exercise the distributed training pipeline, it is recommended to set up a cluster with two or more machines equipped with GPUs either locally or on a cloud instance such as Azure, AWS, or GCP. While the training script we will implement can utilize multiple machines in a cluster, it is not absolutely necessary to have a cluster set up, although it is encouraged.

Now, let's begin!

How to do it...

Since this distributed training setup involves multiple machines, we need a communication interface between the machines and a way to address the individual machines. This is typically done using the existing network infrastructure and IP address:

1. Let's begin by setting up a configuration parameter describing the cluster where we would like to train the models. The following code block is commented out so that you can edit and uncomment based on your cluster setup or leave it commented if you want to simply try it out on a single machine setup:

```
# Uncomment the following lines and fill worker details
# based on your cluster configuration
# tf_config = {
#     "cluster": {"worker": ["1.2.3.4:1111",
                    "localhost:2222"]},
#     "task": {"index": 0, "type": "worker"},
# }
# os.environ["TF_CONFIG"] = json.dumps(tf_config)
```

2. To leverage multi-machine setups, we will use TensorFlow 2.x's `MultiWorkerMirroredStrategy`:

```
strategy = tf.distribute.experimental.
MultiWorkerMirroredStrategy()
```

3. Next, let's declare the basic hyperparameters for the training. Feel free to adjust the batch sizes and the NUM_GPUS values as per your cluster/computer configuration:

```
NUM_GPUS = 2
BS_PER_GPU = 128
NUM_EPOCHS = 60

HEIGHT = 32
WIDTH = 32
```

```
NUM_CHANNELS = 3
NUM_CLASSES = 10
NUM_TRAIN_SAMPLES = 50000


BASE_LEARNING_RATE = 0.1
```

4. To prepare the dataset, let's implement two quick functions for normalizing and augmenting the input images:

```
def normalize(x, y):
    x = tf.image.per_image_standardization(x)
    return x, y
def augmentation(x, y):
    x = tf.image.resize_with_crop_or_pad(x, HEIGHT + 8,
                                         WIDTH + 8)
    x = tf.image.random_crop(x, [HEIGHT, WIDTH,
                                 NUM_CHANNELS])
    x = tf.image.random_flip_left_right(x)
    return x, y
```

5. For the sake of simplicity and faster convergence, we will stick with the CIFAR10 dataset as per the official TensorFlow 2.x sample for training, but feel free to choose a different dataset when you explore. Once you choose the dataset, we can generate the training and the testing sets:

```
(x, y), (x_test, y_test) = \
        keras.datasets.cifar10.load_data()


train_dataset = tf.data.Dataset.from_tensor_slices((x,y))
test_dataset = \
        tf.data.Dataset.from_tensor_slices((x_test, y_test))
```

6. To make the training results reproducible, we will use a fixed random seed to shuffle the dataset:

```
tf.random.set_seed(22)
```

7. We are not ready to generate the training and validation/testing dataset. We will shuffle the dataset using the known and fixed random seed declared in the previous step and apply the augmentation to the training set:

```
train_dataset = (
    train_dataset.map(augmentation)
    .map(normalize)
    .shuffle(NUM_TRAIN_SAMPLES)
    .batch(BS_PER_GPU * NUM_GPUS, drop_remainder=True)
)
```

8. Similarly, we will prepare the test dataset but we do not want to apply random cropping to the test images! So, we will skip the augmentation and use the normalization step for preprocessing:

```
test_dataset = test_dataset.map(normalize).batch(
    BS_PER_GPU * NUM_GPUS, drop_remainder=True
)
```

9. Before we can start training, we need to create an instance of an optimizer and also prepare the input layer. Feel free to use a different optimizer, such as Adam, as per the needs of your task:

```
opt = keras.optimizers.SGD(learning_rate=0.1,
                           momentum=0.9)


input_shape = (HEIGHT, WIDTH, NUM_CHANNELS)
img_input = tf.keras.layers.Input(shape=input_shape)
```

10. Finally, we are ready to construct the model instance within the scope of the `MultiMachineMirroredStrategy`:

```
with strategy.scope():
    model = resnet.resnet56(img_input=img_input,
                            classes=NUM_CLASSES)
    model.compile(
        optimizer=opt,
        loss="sparse_categorical_crossentropy",
        metrics=["sparse_categorical_accuracy"],
    )
```

11. To train the model, we use the simple but powerful Keras API:

```
model.fit(train_dataset, epochs=NUM_EPOCHS)
```

12. Once the model is trained, we easily save, load, and evaluate:

```
# 12.1 Save
model.save(path, save_format="tf")
# 12.2 Load
loaded_model = tf.keras.models.load_model(path)
loaded_model.compile(
    loss=tf.keras.losses.\
        SparseCategoricalCrossentropy(from_logits=True),
    optimizer=tf.keras.optimizers.Adam(),
    metrics=["accuracy"],
)
# 12.3 Evaluate
eval_loss, eval_acc = loaded_model.evaluate(eval_dataset)
```

That completes our recipe implementation! Let's summarize what we implemented and how it works in the next section.

How it works...

For any distributed training runs with TensorFlow 2.x, the `TF_CONFIG` environment variable needs to be set on each of the (virtual) machines on your cluster. These configuration values inform each of the machines about the role and the training information each of the nodes will need to perform its job. You can read more about the details of **TF_CONFIG** configurations used by TensorFlow 2.x's distributed training here: `https://cloud.google.com/ai-platform/training/docs/distributed-training-details`.

We used TensorFlow 2.x's `MultiWorkerMirroredStrategy`, which is a strategy similar to the Mirrored Strategy we used in the previous recipe in this chapter. This strategy is useful for synchronous training across machines with each machine potentially having one or more GPUs. All the variables and computations required for training the model are replicated on each of the worker nodes as in the Mirrored Strategy and, additionally, a distributed collection routine (such as all-reduce) is used to collate results from multiple distributed nodes. The remaining workflow for training, saving the model, loading the model, and evaluating the model remains the same as in our previous recipe.

Ready for the next recipe? Let's do it.

Training Deep RL agents at scale – Multi-GPU PPO agent

RL agents in general require a large number of samples and gradient steps to be trained depending on the complexity of the state, action, and the problem space. With Deep RL, the computational complexity also increases drastically as the deep neural network used by the agent (for Q/value-function representation, for policy representation, or for both) has a lot more operations and parameters that need to be executed and updated, respectively. To speed up the training process, we need the capability to scale our Deep RL agent training to leverage the available compute resources, such as GPUs. This recipe will help you leverage multiple GPUs to train a PPO agent with a deep convolutional neural network policy in a distributed fashion in one of the procedurally generated RL environments using **OpenAI's procgen** library.

Let's get started!

Getting ready

To complete this recipe, you will first need to activate the `tf2rl-cookbook` Python/ conda virtual environment. Make sure to update the environment to match the latest conda environment specification file (`tfrl-cookbook.yml`) in the cookbook's code repo. Although not required, it is recommended to use a machine with two or more GPUs to execute this recipe.

Now, let's begin!

How to do it...

We will implement a complete recipe to allow configurable training of a PPO agent with a deep convolutional neural network policy in a distributed fashion. Let's start implementing it step by step:

1. We will begin by importing the necessary modules for our recipe:

```python
import argparse
import os
from datetime import datetime

import gym
import gym.wrappers
import numpy as np
import tensorflow as tf
from tensorflow.keras.layers import (
    Conv2D,
    Dense,
    Dropout,
    Flatten,
    Input,
    MaxPool2D,
)
```

2. We will be using the `procgen` environments from OpenAI. Let's import that as well:

```
import procgen  # Import & register procgen Gym envs
```

3. In order to make this recipe easy to configure and run, let's add support for command-line arguments with useful configuration flags:

```
parser = argparse.ArgumentParser(prog="TFRL-Cookbook-Ch9-
Distributed-RL-Agent")
parser.add_argument("--env", default="procgen:procgen-
coinrun-v0")
parser.add_argument("--update-freq", type=int,
default=16)
parser.add_argument("--epochs", type=int, default=3)
parser.add_argument("--actor-lr", type=float,
default=1e-4)
parser.add_argument("--critic-lr", type=float,
default=1e-4)
parser.add_argument("--clip-ratio", type=float,
default=0.1)
parser.add_argument("--gae-lambda", type=float,
default=0.95)
parser.add_argument("--gamma", type=float, default=0.99)
parser.add_argument("--logdir", default="logs")

args = parser.parse_args()
```

4. Let's use a TensorBoard summary writer for logging:

```
logdir = os.path.join(
    args.logdir, parser.prog, args.env, \
    datetime.now().strftime("%Y%m%d-%H%M%S")
)
print(f"Saving training logs to:{logdir}")
writer = tf.summary.create_file_writer(logdir)
```

5. We will first implement the `Actor` class in the following several steps, starting with the `__init__` method. You will notice that we need to instantiate the models within the context of the execution strategy:

```python
class Actor:
    def __init__(self, state_dim, action_dim,
    execution_strategy):
        self.state_dim = state_dim
        self.action_dim = action_dim
        self.execution_strategy = execution_strategy
        with self.execution_strategy.scope():
            self.weight_initializer = \
                tf.keras.initializers.he_normal()
            self.model = self.nn_model()
            self.model.summary()  # Print a summary of
            # the Actor model
            self.opt = \
                tf.keras.optimizers.Nadam(args.actor_lr)
```

6. For the Actor's policy network model, we will implement a deep convolutional neural network comprising of multiple `Conv2D` and `MaxPool2D` layers. We will start the implementation in this step and finish in the following few steps:

```python
def nn_model(self):
    obs_input = Input(self.state_dim)
    conv1 = Conv2D(
        filters=64,
        kernel_size=(3, 3),
        strides=(1, 1),
        padding="same",
        input_shape=self.state_dim,
        data_format="channels_last",
        activation="relu",
    )(obs_input)
    pool1 = MaxPool2D(pool_size=(3, 3), \
                        strides=1)(conv1)
```

7. We will add more Conv2D – Pool2D layers to stack up the processing layers depending on the needs for the task. In this recipe, we will be training policies for the procgen environment, which is somewhat visually rich, so we will stack a few more layers:

```
conv2 = Conv2D(
    filters=32,
    kernel_size=(3, 3),
    strides=(1, 1),
    padding="valid",
    activation="relu",
)(pool1)
pool2 = MaxPool2D(pool_size=(3, 3), strides=1)\
            (conv2)
conv3 = Conv2D(
    filters=16,
    kernel_size=(3, 3),
    strides=(1, 1),
    padding="valid",
    activation="relu",
)(pool2)
pool3 = MaxPool2D(pool_size=(3, 3), strides=1)\
            (conv3)
conv4 = Conv2D(
    filters=8,
    kernel_size=(3, 3),
    strides=(1, 1),
    padding="valid",
    activation="relu",
)(pool3)
pool4 = MaxPool2D(pool_size=(3, 3), strides=1)\
            (conv4)
```

8. Now, we can use a flattening layer and prepare the output heads for the policy network:

```
flat = Flatten()(pool4)
dense1 = Dense(
    16, activation="relu", \
        kernel_initializer=self.weight_initializer
)(flat)
dropout1 = Dropout(0.3)(dense1)
dense2 = Dense(
    8, activation="relu", \
        kernel_initializer=self.weight_initializer
)(dropout1)
dropout2 = Dropout(0.3)(dense2)
```

9. As the final step for building the neural model for the policy network, we will create the output layer and return a Keras model:

```
output_discrete_action = Dense(
    self.action_dim,
    activation="softmax",
    kernel_initializer=self.weight_initializer,
)(dropout2)
return tf.keras.models.Model(
    inputs=obs_input,
    outputs = output_discrete_action,
    name="Actor")
```

10. With the model we have defined in the previous steps, we can start processing a state/observation image input and produce the logits (unnormalized probabilities) and the action that the Actor would take. Let's implement a method to do that:

```
def get_action(self, state):
    # Convert [Image] to np.array(np.adarray)
    state_np = np.array([np.array(s) for s in state])
    if len(state_np.shape) == 3:
        # Convert (w, h, c) to (1, w, h, c)
        state_np = np.expand_dims(state_np, 0)
```

```
logits = self.model.predict(state_np)
# shape: (batch_size, self.action_dim)
action = np.random.choice(self.action_dim,
                            p=logits[0])
# 1 Action per instance of env; Env expects:
# (num_instances, actions)
# action = (action,)
return logits, action
```

11. Next, to compute the surrogate loss to drive the learning, we will implement the compute_loss method:

```
def compute_loss(self, old_policy, new_policy,
actions, gaes):
    log_old_policy = tf.math.log(tf.reduce_sum(
                        old_policy * actions))
    log_old_policy = tf.stop_gradient(log_old_policy)
    log_new_policy = tf.math.log(tf.reduce_sum(
                        new_policy * actions))
    # Avoid INF in exp by setting 80 as the upper
    # bound since,
    # tf.exp(x) for x>88 yeilds NaN (float32)
    ratio = tf.exp(
        tf.minimum(log_new_policy - \
                    tf.stop_gradient(log_old_policy),\
                    80)
    )
    clipped_ratio = tf.clip_by_value(
        ratio, 1.0 - args.clip_ratio, 1.0 + \
        args.clip_ratio
    )
    gaes = tf.stop_gradient(gaes)
    surrogate = -tf.minimum(ratio * gaes, \
                            clipped_ratio * gaes)
    return tf.reduce_mean(surrogate)
```

12. Next up is a core method that ties all the methods together to perform the training. Note that this is the train method per replica, and we will use it in our distributed training method, which will follow in the next steps:

```
def train(self, old_policy, states, actions, gaes):
    actions = tf.one_hot(actions, self.action_dim)
    # One-hot encoding
    actions = tf.reshape(actions, [-1, \
                            self.action_dim])
    # Add batch dimension
    actions = tf.cast(actions, tf.float64)
    with tf.GradientTape() as tape:
        logits = self.model(states, training=True)
        loss = self.compute_loss(old_policy, logits,
                            actions, gaes)
    grads = tape.gradient(loss,
                        self.model.trainable_variables)
    self.opt.apply_gradients(zip(grads,
                        self.model.trainable_variables))
    return loss
```

13. To implement the distributed training method, we will make use of the `tf.function` decorator to implement a TensorFlow 2.x function:

```
@tf.function
def train_distributed(self, old_policy, states,
                        actions, gaes):
    per_replica_losses = self.execution_strategy.run(
        self.train, args=(old_policy, states,
                        actions, gaes))
    return self.execution_strategy.reduce(
        tf.distribute.ReduceOp.SUM, \
            per_replica_losses, axis=None)
```

14. That completes our `Actor` class implementation, and we will now start our implementation of the `Critic` class:

```python
class Critic:
    def __init__(self, state_dim, execution_strategy):
        self.state_dim = state_dim
        self.execution_strategy = execution_strategy
        with self.execution_strategy.scope():
            self.weight_initializer = \
                tf.keras.initializers.he_normal()
            self.model = self.nn_model()
            self.model.summary()
            # Print a summary of the Critic model
            self.opt = \
                tf.keras.optimizers.Nadam(args.critic_lr)
```

15. You must have noticed that we are creating the Critic's value-function model instance within the scope of the execution strategy to support distributed training. We will now start implementing the Critic's neural network model in the following few steps:

```python
def nn_model(self):
    obs_input = Input(self.state_dim)
    conv1 = Conv2D(
        filters=64,
        kernel_size=(3, 3),
        strides=(1, 1),
        padding="same",
        input_shape=self.state_dim,
        data_format="channels_last",
        activation="relu",
    )(obs_input)
    pool1 = MaxPool2D(pool_size=(3, 3), strides=2)\
            (conv1)
```

16. Like our Actor's model, we will have similar layering of Conv2D-MaxPool2D layers followed by flattening layers with dropout:

```
conv2 = Conv2D(filters=32, kernel_size=(3, 3),
        strides=(1, 1),
        padding="valid", activation="relu",)(pool1)
pool2 = MaxPool2D(pool_size=(3, 3), strides=2)\
        (conv2)
conv3 = Conv2D(filters=16,
        kernel_size=(3, 3), strides=(1, 1),
        padding="valid", activation="relu",)(pool2)
pool3 = MaxPool2D(pool_size=(3, 3), strides=1)\
        (conv3)
conv4 = Conv2D(filters=8, kernel_size=(3, 3),
        strides=(1, 1), padding="valid",
        activation="relu",)(pool3)
pool4 = MaxPool2D(pool_size=(3, 3), strides=1)\
        (conv4)
flat = Flatten()(pool4)
dense1 = Dense(16, activation="relu",
                kernel_initializer =\
                self.weight_initializer)\
            (flat)
dropout1 = Dropout(0.3)(dense1)
dense2 = Dense(8, activation="relu",
                kernel_initializer = \
                self.weight_initializer)\
            (dropout1)
dropout2 = Dropout(0.3)(dense2)
```

17. We will add the value output head and return the model as a Keras model to complete our Critic's neural network model:

```
value = Dense(
    1, activation="linear",
    kernel_initializer=self.weight_initializer)\
    (dropout2)
```

```
        return tf.keras.models.Model(inputs=obs_input, \
                                     outputs=value, \
                                     name="Critic")
```

18. As you may recall, the Critic's loss is the mean squared error between the predicted temporal-difference target and the actual temporal-difference targets. Let's implement a method to compute the loss:

```
def compute_loss(self, v_pred, td_targets):
    mse = tf.keras.losses.MeanSquaredError(
            reduction=tf.keras.losses.Reduction.SUM)
    return mse(td_targets, v_pred)
```

19. Similar to our Actor implementation, we will implement a per-replica `train` method and then use it in a later step for distributed training:

```
def train(self, states, td_targets):
    with tf.GradientTape() as tape:
        v_pred = self.model(states, training=True)
        # assert v_pred.shape == td_targets.shape
        loss = self.compute_loss(v_pred, \
                        tf.stop_gradient(td_targets))
    grads = tape.gradient(loss, \
                    self.model.trainable_variables)
    self.opt.apply_gradients(zip(grads, \
                    self.model.trainable_variables))
    return loss
```

20. We will now finalize our `Critic` class implementation by implementing the `train_distributed` method that enables distributed training:

```
@tf.function
def train_distributed(self, states, td_targets):
    per_replica_losses = self.execution_strategy.run(
        self.train, args=(states, td_targets)
```

```
        )
        return self.execution_strategy.reduce(
            tf.distribute.ReduceOp.SUM, \
            per_replica_losses, axis=None
        )
```

21. With our `Actor` and `Critic` classes implemented, we can start our distributed `PPOAgent` implementation. We will implement the `PPOAgent` class in the following several steps. Let's begin with the `__init__` method:

```
class PPOAgent:
    def __init__(self, env):
        """Distributed PPO Agent for image observations
        and discrete action-space Gym envs

        Args:
            env (gym.Env): OpenAI Gym I/O compatible RL
            environment with discrete action space
        """
        self.env = env
        self.state_dim = self.env.observation_space.shape
        self.action_dim = self.env.action_space.n
        # Create a Distributed execution strategy
        self.distributed_execution_strategy = \
                    tf.distribute.MirroredStrategy()
        print(f"Number of devices: {self.\
                distributed_execution_strategy.\
                num_replicas_in_sync}")
        # Create Actor & Critic networks under the
        # distributed execution strategy scope
        with self.distributed_execution_strategy.scope():
            self.actor = Actor(self.state_dim,
                                self.action_dim,
                                tf.distribute.get_strategy())
            self.critic = Critic(self.state_dim,
                                tf.distribute.get_strategy())
```

22. Next, we will implement a method to calculate the target for the **generalized advantage estimate (GAE)**:

```python
def gae_target(self, rewards, v_values, next_v_value,
done):
    n_step_targets = np.zeros_like(rewards)
    gae = np.zeros_like(rewards)
    gae_cumulative = 0
    forward_val = 0

    if not done:
        forward_val = next_v_value

    for k in reversed(range(0, len(rewards))):
        delta = rewards[k] + args.gamma * \
            forward_val - v_values[k]
        gae_cumulative = args.gamma * \
            args.gae_lambda * gae_cumulative + delta
        gae[k] = gae_cumulative
        forward_val = v_values[k]
        n_step_targets[k] = gae[k] + v_values[k]
    return gae, n_step_targets
```

23. We are all set to start our main `train(...)` method. We will split the implementation of this method into the following few steps. Let's set up the scope, start the outer loop, and initialize varaibles:

```python
def train(self, max_episodes=1000):

    with self.distributed_execution_strategy.scope():
        with writer.as_default():
            for ep in range(max_episodes):
                state_batch = []
                action_batch = []
                reward_batch = []
                old_policy_batch = []
```

```
episode_reward, done = 0, False

state = self.env.reset()
prev_state = state
step_num = 0
```

24. Now, we can start the loop that needs to be executed for each episode until the episode is done:

```
while not done:
    self.env.render()
    logits, action = \
            self.actor.get_action(state)

    next_state, reward, dones, _ = \
                self.env.step(action)
    step_num += 1
    print(f"ep#:{ep} step#:{step_num}
            step_rew:{reward} \
            action:{action} \
            dones:{dones}",end="\r",)
    done = np.all(dones)
    if done:
        next_state = prev_state
    else:
        prev_state = next_state

    state_batch.append(state)
    action_batch.append(action)
    reward_batch.append(
                (reward + 8) / 8)
    old_policy_batch.append(logits)
```

25. Within each episode, if we have reached `update_freq` or just reached an end state, we need to compute the GAEs and TD targets. Let's add the code for that:

```python
if len(state_batch) >= \
args.update_freq or done:
    states = np.array(
        [state.squeeze() for \
        state in state_batch])
    actions = \
        np.array(action_batch)
    rewards = \
        np.array(reward_batch)
    old_policies = np.array(
        [old_pi.squeeze() for \
        old_pi in old_policy_batch])

    v_values = self.critic.\
        model.predict(states)
    next_v_value = self.critic.\
        model.predict(
            np.expand_dims(
                next_state, 0))

    gaes, td_targets = \
        self.gae_target(
            rewards, v_values,
            next_v_value, done)
    actor_losses, critic_losses=\
        [], []
```

26. Within the same execution context, we need to train the `Actor` and the `Critic`:

```python
for epoch in range(args.\
epochs):
    actor_loss = self.actor.\
        train_distributed(
            old_policies,
```

```
                            states, actions,
                            gaes)
                    actor_losses.\
                        append(actor_loss)
                    critic_loss = self.\
                        critic.train_distributed(
                        states, td_targets)
                    critic_losses.\
                        append(critic_loss)
                # Plot mean actor & critic
                # losses on every update
                tf.summary.scalar(
                    "actor_loss",
                        np.mean(actor_losses),
                        step=ep)
                tf.summary.scalar(
                    "critic_loss",
                        np.mean(critic_losses),
                        step=ep)
```

27. Finally, we need to reset the tracking variables and update our episode reward values:

```
                state_batch = []
                action_batch = []
                reward_batch = []
                old_policy_batch = []

            episode_reward += reward
            state = next_state
```

28. With that, our distributed **PPOAgent** implementation is complete! We will implement our `main` method to finalize our recipe:

```
if __name__ == "__main__":
    env_name = "procgen:procgen-coinrun-v0"
    env = gym.make(env_name, render_mode="rgb_array")
    env = gym.wrappers.Monitor(env=env,
                            directory="./videos", force=True)
    agent = PPOAgent(env)
    agent.train()
```

That's it for the recipe! Hope you enjoyed cooking it up. You can execute the recipe and watch the progress using the TensorBoard logs to see the training speedup you get with a greater number of GPUs!

Let's recap what we accomplished and how the recipe works in the next section.

How it works...

We implemented `Actor` and `Critic` classes where the Actor used a deep convolutional neural network for the policy representation and the Critic utilized a similar deep convolutional neural network for its value function representation. Both these models were instantiated under the scope of the distributed execution strategy using the `self.execution_strategy.scope()` construct.

The procgen environments, such as coinrun, fruitbot, jumper, leaper, maze, and others, are visually (relatively) rich environments and therefore require convolutional layers that are relatively deep to process the visual observations. We therefore used a deep CNN model for the policy network of the Actor. For distributed training using multiple replicas on multiple GPUs, we first implemented a single-replica training method (train) and then used `Tensorflow.function` to run across replicas and reduce the results to arrive at the total loss.

Finally, while training our PPO agent in the distributed setting, we performed all the training operations within the scope of the distributed execution strategy by using Python's `with` statement for context management like this: `with self.distributed_execution_strategy.scope()`.

It's time to move on to the next recipe!

Building blocks for distributed Deep Reinforcement Learning for accelerated training

The previous recipes in this chapter discussed how you could scale your Deep RL training using TensorFlow 2.x's distributed execution APIs. While it was straightforward after understanding the concepts and the implementation style, training Deep RL agents with more advanced architectures such as Impala and R2D2 requires RL building blocks such as distributed parameter servers and distributed experience replay. This chapter will walk through the implementation of such building blocks for distributed RL training. We will be using the Ray distributed computing framework to implement our building blocks.

Let's get started!

Getting ready

To complete this recipe, you will first need to activate the tf2rl-cookbook Python/conda virtual environment. Make sure to update the environment to match the latest conda environment specification file (tfrl-cookbook.yml) in the cookbook's code repo. To test the building blocks we build in this recipe, we will be using the sac_agent_base module based on our SAC agent implemented in one of the book's earlier recipes. If the following import statements run without issues, you are ready to start:

```
import pickle
import sys


import fire
import gym
import numpy as np
import ray


if "." not in sys.path:
    sys.path.insert(0, ".")
from sac_agent_base import SAC
```

Now, let's begin!

How to do it...

We will implement the building blocks one by one, starting with the distributed parameter server:

1. The `ParameterServer` class is a simple store for sharing the neural network parameters or weights between workers in a distributed training setting. We will implement the class as a Ray's remote Actor:

```
@ray.remote
class ParameterServer(object):
    def __init__(self, weights):
        values = [value.copy() for value in weights]
        self.weights = values

    def push(self, weights):
        values = [value.copy() for value in weights]
        self.weights = values
    def pull(self):
        return self.weights

    def get_weights(self):
        return self.weights
```

2. Let's also add a method to save the weights to the disk:

```
# save weights to disk
def save_weights(self, name):
    with open(name + "weights.pkl", "wb") as pkl:
        pickle.dump(self.weights, pkl)
    print(f"Weights saved to {name +
                             'weights.pkl'}.")
```

3. As the next building block, we will implement the `ReplayBuffer`, which can be used by a distributed set of agents. We will start the implementation in this step and continue in the next several steps:

```
@ray.remote
class ReplayBuffer:
```

```
"""
A simple FIFO experience replay buffer for RL Agents
"""
def __init__(self, obs_shape, action_shape, size):
    self.cur_states = np.zeros([size, obs_shape[0]],
                                dtype=np.float32)
    self.actions = np.zeros([size, action_shape[0]],
                            dtype=np.float32)
    self.rewards = np.zeros(size, dtype=np.float32)
    self.next_states = np.zeros([size, obs_shape[0]],
                                dtype=np.float32)
    self.dones = np.zeros(size, dtype=np.float32)
    self.idx, self.size, self.max_size = 0, 0, size
    self.rollout_steps = 0
```

4. Next, we will implement a method to store new experiences in the replay buffer:

```
def store(self, obs, act, rew, next_obs, done):
    self.cur_states[self.idx] = np.squeeze(obs)
    self.actions[self.idx] = np.squeeze(act)
    self.rewards[self.idx] = np.squeeze(rew)
    self.next_states[self.idx] = np.squeeze(next_obs)
    self.dones[self.idx] = done
    self.idx = (self.idx + 1) % self.max_size
    self.size = min(self.size + 1, self.max_size)
    self.rollout_steps += 1
```

5. To sample a batch of experience data from the replay buffer, we will implement a method that randomly samples from the replay buffer and returns a dictionary containing the sampled experience data:

```
def sample_batch(self, batch_size=32):
    idxs = np.random.randint(0, self.size,
                             size=batch_size)
    return dict(
        cur_states=self.cur_states[idxs],
```

```
            actions=self.actions[idxs],
            rewards=self.rewards[idxs],
            next_states=self.next_states[idxs],
            dones=self.dones[idxs])
```

6. That completes our `ReplayBuffer` class implementation. We will now start implementing a method to roll out, which essentially collects experiences in an RL environment using an exploration policy with parameters pulled from the distributed parameter server object and stores the collected experience in the distributed replay buffer. We will start our implementation in this step and complete the `rollout` method implementation in the following steps:

```
@ray.remote
def rollout(ps, replay_buffer, config):
    """Collect experience using an exploration policy"""
    env = gym.make(config["env"])
    obs, reward, done, ep_ret, ep_len = env.reset(), 0, \
                                        False, 0, 0
    total_steps = config["steps_per_epoch"] * \
                  config["epochs"]

    agent = SAC(env.observation_space.shape, \
                env.action_space)
    weights = ray.get(ps.pull.remote())
    target_weights = agent.actor.get_weights()
    for i in range(len(target_weights)):
    # set tau% of target model to be new weights
        target_weights[i] = weights[i]
    agent.actor.set_weights(target_weights)
```

7. With the agent intialized and loaded and the environment instance(s) ready, we can start our experience-gathering loop:

```
        for step in range(total_steps):
            if step > config["random_exploration_steps"]:
                # Use Agent's policy for exploration after
```

```
        `random_exploration_steps`
        a = agent.act(obs)
    else:  # Use a uniform random exploration policy
        a = env.action_space.sample()


    next_obs, reward, done, _ = env.step(a)
    print(f"Step#:{step} reward:{reward} \
            done:{done}")
    ep_ret += reward
    ep_len += 1
```

8. Let's handle the case when a `max_ep_len` is configured to indicate the maximum length of the episode and then store the collected experience in the distributed replay buffer:

```
    done = False if ep_len == config["max_ep_len"]\
            else done
    # Store experience to replay buffer
    replay_buffer.store.remote(obs, a, reward,
                        next_obs, done)
```

9. Finally, at the end of the episode, sync the weights of the behavior policy using the parameter server:

```
    obs = next_obs


    if done or (ep_len == config["max_ep_len"]):
        """
        Perform parameter sync at the end of the
        trajectory.
        """
        obs, reward, done, ep_ret, ep_len = \
                    env.reset(), 0, False, 0, 0
        weights = ray.get(ps.pull.remote())
        agent.actor.set_weights(weights)
```

10. That completes the implementation of the `rollout` method and we can now implement a `train` method that runs the train loop:

```python
@ray.remote(num_gpus=1, max_calls=1)
def train(ps, replay_buffer, config):
    agent = SAC(config["obs_shape"], \
                config["action_space"])
    weights = ray.get(ps.pull.remote())
    agent.actor.set_weights(weights)
    train_step = 1
    while True:

        agent.train_with_distributed_replay_memory(
            ray.get(replay_buffer.sample_batch.remote())
        )

        if train_step % config["worker_update_freq"] == 0:
            weights = agent.actor.get_weights()
            ps.push.remote(weights)
        train_step += 1
```

11. The final module in our recipe is the `main` function, which puts together all the building blocks we have built so far in this recipe and exercises them. We will begin the implementation in this step and finish it in the remaining steps. Let's start with the `main` function argument list and capture the arguments in a config dictionary:

```python
def main(
    env="MountainCarContinuous-v0",
    epochs=1000,
    steps_per_epoch=5000,
    replay_size=100000,
    random_exploration_steps=1000,
    max_ep_len=1000,
    num_workers=4,
    num_learners=1,
    worker_update_freq=500,
):
```

```
config = {
    "env": env,
    "epochs": epochs,
    "steps_per_epoch": steps_per_epoch,
    "max_ep_len": max_ep_len,
    "replay_size": replay_size,
    "random_exploration_steps": \
            random_exploration_steps,
    "num_workers": num_workers,
    "num_learners": num_learners,
    "worker_update_freq": worker_update_freq,
}
```

12. Next, let's create an instance of the desired environment, obtain the state and observation space, initialize ray, and also initialize a Stochastic Actor-Critic agent. Note that we are initializing a single-node ray cluster but feel free to initialize ray with a cluster of nodes (local or in the cloud):

```
env = gym.make(config["env"])
config["obs_shape"] = env.observation_space.shape
config["action_space"] = env.action_space


ray.init()


agent = SAC(config["obs_shape"], \
            config["action_space"])
```

13. In this step, we will initialize an instance of the `ParameterServer` class and an instance of the `ReplayBuffer` class:

```
params_server = \
        ParameterServer.remote(agent.actor.get_weights())


replay_buffer = ReplayBuffer.remote(
    config["obs_shape"], \
    config["action_space"].shape, \
```

```
        config["replay_size"]
    )
```

14. We are now ready to exercise the building blocks we have built. We will first launch a series of rollout tasks based on the number of workers specified as a configuration argument that will launch the rollout process on the distributed ray cluster:

```
task_rollout = [
    rollout.remote(params_server, replay_buffer,
                    config)
    for i in range(config["num_workers"])
]
```

The rollout task will launch the remote tasks that will populate the replay buffer with the gathered experience. The above line will return immediately even though the rollout tasks will take time to complete because of the asynchronous function call.

15. Next, we will launch a configurable number of learners that run the distributed training task on the ray cluster:

```
task_train = [
    train.remote(params_server, replay_buffer,
                    config)
    for i in range(config["num_learners"])
]
```

The above statement will launch the remote training process and return immediately even though the train function on the learners will take time to complete.

```
We will wait for the tasks to complete on the main thread
before exiting:
    ray.wait(task_rollout)
    ray.wait(task_train)
```

16. Finally, let's define our entry point. We will use the Python Fire library to expose our `main` function, and its arguments to look like an executable supporting command-line argument:

```
if __name__ == "__main__":
    fire.Fire(main)
```

With the preceding entry point, the script can be configured and launched from the command line. An example is provided here for your reference:

```
(tfrl-cookbook)praveen@dev-cluster:~/tfrl-cookbook$python
4_building_blocks_for_distributed_rl_using_ray.py main
--env="MountaincarContinuous-v0" --num_workers=8 --num_
learners=3
```

That completes our implementation! Let's briefly discuss how it works in the next section.

How it works...

We built a distributed `ParameterServer`, `ReplayBuffer`, rollout worker, and learner processes. These building blocks are crucial for training distributed RL agents. We utilized Ray as the framework for distributed computing.

After implementing the building blocks and the tasks, in the `main` function, we launched the two asynchronous, distributed tasks on the ray cluster. The `task_rollout` launched a (configurable) number of rollout workers and the `task_train` launched a (configurable) number of learners. Both the tasks run on the ray cluster asynchronously in a distributed manner. The rollout workers pull the latest weights from the parameter server and gather and store experiences in the replay memory buffer while, simultaneously, the learners train using batches of experiences sampled from the replay memory and push the updated (and potentially improved) set of parameters to the parameter server.

It's time to move on to the next and final recipe of this chapter!

Large-scale Deep RL agent training using Ray, Tune, and RLLib

In the previous recipe, we got a flavor of how to implement distributed RL agent training routines from scratch. Since most of the components used as building blocks have become a standard way of building Deep RL training infrastructure, we can leverage an existing library that maintains a high-quality implementation of such building blocks. Fortunately, with our choice of ray as the framework for distributed computing, we are in a good place. Tune and RLLib are two libraries built on top of ray, and are available together with Ray, that provide highly scalable hyperparameter tuning (Tune) and RL training (RLLib). This recipe will provide a curated set of steps to get you acquainted with ray, Tune, and RLLib so that you can utilize them to scale your Deep RL training routines. In addition to the recipe discussed here in the text, the cookbook's code repository for this chapter will have a handful of additional recipes for you.

Let's get started!

Getting ready

To complete this recipe, you will first need to activate the tf2rl-cookbook Python/ conda virtual environment. Make sure to update the environment to match the latest conda environment specification file (tfrl-cookbook.yml) in the cookbook's code repo. Ray, Tune, and RLLib will be installed in your tfrl-cookbook conda environment when you use the provided conda YAML spec for the environment. If you want to install Tune and RLLib in a different environment, the easiest way is to install it using the following command:

```
pip install ray[tune,rllib]
```

Now, let's begin!

How to do it...

We will start with quick and basic commands and recipes to launch training on ray clusters using Tune and RLLib and progressively customize the training pipeline to provide you with a useful recipe:

1. Launching typical training of RL agents in OpenAI Gym environments is as easy as specifying the algorithm name and the environment name. For example, to train a PPO agent in the CartPole-v4 Gym environment, all you need to execute is the following command:

   ```
   (tfrl-cookbook) praveen@dev-cluster:~/tfrl-cookbook$rllib
   train -run PPO -env "CartPole-v4" --eager
   ```

   Note that the `--eager` flag is also specified, which forces RLLib to use eager execution (the default mode of execution in TensorFlow 2.x).

2. Let's try to train a PPO agent in the `coinrun procgen` environment, like in one of our previous recipes:

   ```
   (tfrl-cookbook) praveen@dev-cluster:~/tfrl-cookbook$rllib
   train --run PPO --env "procgen:procgen-coinrun-v0"
   --eager
   ```

   You will notice that the preceding command fails with the following (shortened) error:

   ```
       ValueError: No default configuration for obs shape
   [64, 64, 3], you must specify `conv_filters` manually as
   a model option. Default configurations are only available
   for inputs of shape [42, 42, K] and [84, 84, K]. You may
   alternatively want to use a custom model or preprocessor.
   ```

   This is because, as the error states, RLLib by default supports observations of shapes (42, 42, k) or (84, 84, k). Observations of other shapes will need a custom model or a preprocessor. In the next few steps, we will see how we can implement a custom neural network model implemented using the TensorFlow 2.x Keras API, which can be used with ray RLLib.

3. We will start our custom model implementation (`custom_model.py`) in this step and complete it in the following few steps. In this step, let's import the necessary modules and also implement a helper method to return a Conv2D layer with a certain filter depth:

   ```
   from ray.rllib.models.tf.tf_modelv2 import TFModelV2
   import tensorflow as tf
   ```

```python
def conv_layer(depth, name):
    return tf.keras.layers.Conv2D(
        filters=depth, kernel_size=3, strides=1, \
        padding="same", name=name
    )
```

4. Next, let's implement a helper method to build and return a simple residual block:

```python
def residual_block(x, depth, prefix):
    inputs = x
    assert inputs.get_shape()[-1].value == depth
    x = tf.keras.layers.ReLU()(x)
    x = conv_layer(depth, name=prefix + "_conv0")(x)
    x = tf.keras.layers.ReLU()(x)
    x = conv_layer(depth, name=prefix + "_conv1")(x)
    return x + inputs
```

5. Let's implement another handy function to construct multiple residual block sequences:

```python
def conv_sequence(x, depth, prefix):
    x = conv_layer(depth, prefix + "_conv")(x)
    x = tf.keras.layers.MaxPool2D(pool_size=3, \
                                  strides=2,\
                                  padding="same")(x)
    x = residual_block(x, depth, prefix=prefix + \
                       "_block0")
    x = residual_block(x, depth, prefix=prefix + \
                       "_block1")
    return x
```

6. We can now start implementing the CustomModel class as a subclass of the TFModelV2 base class provided by RLLib to make it easy to integrate with RLLib:

```python
class CustomModel(TFModelV2):
    """Deep residual network that produces logits for
```

```
        policy and value for value-function;
    Based on architecture used in IMPALA paper:https://
        arxiv.org/abs/1802.01561"""


    def __init__(self, obs_space, action_space,
    num_outputs, model_config, name):
        super().__init__(obs_space, action_space, \
                        num_outputs, model_config, name)
        depths = [16, 32, 32]
        inputs = tf.keras.layers.Input(
                        shape=obs_space.shape,
                        name="observations")
        scaled_inputs = tf.cast(inputs,
                            tf.float32) / 255.0
        x = scaled_inputs
        for i, depth in enumerate(depths):
            x = conv_sequence(x, depth, prefix=f"seq{i}")
        x = tf.keras.layers.Flatten()(x)
        x = tf.keras.layers.ReLU()(x)
        x = tf.keras.layers.Dense(units=256,
                            activation="relu",
                            name="hidden")(x)
        logits = tf.keras.layers.Dense(units=num_outputs,
                            name="pi")(x)
        value = tf.keras.layers.Dense(units=1,
                            name="vf")(x)
        self.base_model = tf.keras.Model(inputs,
                            [logits, value])
        self.register_variables(
                            self.base_model.variables)
```

7. After the __init__ method, we need to implement the forward method as it is not implemented by the base class (TFModelV2) but is necessary:

```
    def forward(self, input_dict, state, seq_lens):
        # explicit cast to float32 needed in eager
```

```
obs = tf.cast(input_dict["obs"], tf.float32)
logits, self._value = self.base_model(obs)
return logits, state
```

8. We will also implement a one-line method to reshape the value function output:

```
def value_function(self):
    return tf.reshape(self._value, [-1])
```

With that, our `CustomModel` implementation is complete and is ready to use!

9. We will implement a solution (`5.1_training_using_tune_run.py`) using ray, Tune, and RLLib's Python API so that you also utilize the model in addition to their command-line usage. Let's split the implementation into two steps. In this step, we will import the necessary modules and initialize ray:

```
import ray
import sys
from ray import tune
from ray.rllib.models import ModelCatalog


if not "." in sys.path:
    sys.path.insert(0, ".")
from custom_model import CustomModel
ray.init()  # Can also initialize a cluster with multiple
#nodes here using the cluster head node's IP
```

10. In this step, we will register our custom model in RLLib's `ModelCatlog` and then use it to train a PPO agent with a custom set of parameters including the `framework` parameter that forces RLLib to use TensorFlow 2. We will also shut down ray at the end of the script:

```
# Register custom-model in ModelCatalog
ModelCatalog.register_custom_model("CustomCNN",
                                    CustomModel)


experiment_analysis = tune.run(
    "PPO",
    config={
```

```
        "env": "procgen:procgen-coinrun-v0",
        "num_gpus": 0,
        "num_workers": 2,
        "model": {"custom_model": "CustomCNN"},
        "framework": "tf2",
        "log_level": "INFO",
    },
    local_dir="ray_results",  # store experiment results
    #  in this dir
)
ray.shutdown()
```

11. We will look at another quick recipe (5_2_custom_training_using_tune.
    py) to customize the training loop. We will split the implementation into the
    following few steps to keep it simple. In this step, we will import the necessary
    libraries and initialize ray:

```
import sys

import ray
import ray.rllib.agents.impala as impala
from ray.tune.logger import pretty_print
from ray.rllib.models import ModelCatalog

if not "." in sys.path:
    sys.path.insert(0, ".")
from custom_model import CustomModel
ray.init()  # You can also initialize a multi-node ray
# cluster here
```

12. Now, let's register our custom model with RLLib's ModelCatalog and configure
    the **IMPALA agent**. We could very well use any other RLLib support agents, such as
    PPO or SAC:

```
# Register custom-model in ModelCatalog
ModelCatalog.register_custom_model("CustomCNN",
```

```
                                                CustomModel)

config = impala.DEFAULT_CONFIG.copy()
config["num_gpus"] = 0
config["num_workers"] = 1
config["model"]["custom_model"] = "CustomCNN"
config["log_level"] = "INFO"
config["framework"] = "tf2"
trainer = impala.ImpalaTrainer(config=config,
                        env="procgen:procgen-coinrun-v0")
```

13. We can now implement our custom training loop and include any steps in the loop as we desire. We will keep the example loop simple by simply performing a training step and saving the agent's model every n(100) epochs:

```
for step in range(1000):
    # Custom training loop
    result = trainer.train()
    print(pretty_print(result))

    if step % 100 == 0:
        checkpoint = trainer.save()
        print("checkpoint saved at", checkpoint
```

14. Note that we could continue to train the agent using the saved checkpoint and using the simpler ray tune's run API as shown here as an example:

```
# Restore agent from a checkpoint and start a new
# training run with a different config
config["lr"] =  ray.tune.grid_search([0.01, 0.001])"]
ray.tune.run(trainer, config=config, restore=checkpoint)
```

15. Finally, let's shut down ray to free up system resources:

```
ray.shutdown()
```

That completes this recipe! In the next section, let's recap what we discussed in this recipe.

How it works...

We identified one of the common limitations with the simple but limited command-line interface of ray RLLib. We also discussed a solution to overcome the failure in step 2 where a custom model was needed to use RLLib's PPO agent training and implemented it in steps 9 and 10.

While the solution discussed in steps 9 and 10 looks elegant, it may not provide all the customization knobs you are looking for or are familiar with. For example, it abstracts away the basic RL loop that steps through the environment. We implemented another quick recipe starting from step 11 that allows the customization of the training loop. In step 12, we saw how we can register our custom model and use it with the IMPALA agent – which is a scalable, distributed Deep RL agent based on IMPortance-weighted Actor-Learner Architecture. IMPALA agent actors communicate sequences of states, actions, and rewards to a centralized learner where batch gradient updates take place, in contrast to the (asynchronous) Actor-Critic-based agents where the gradients are communicated to a centralized parameter server.

For more information on Tune, you can refer to the Tune user guide and configuration documentation at `https://docs.ray.io/en/master/tune/user-guide.html`.

For more information on RLLib training APIs and configuration documentation, you can refer to `https://docs.ray.io/en/master/rllib-training.html`.

That completes the recipe and the chapter! Hope you feel empowered with the new skills and knowledge you have gained to speed up your Deep RL agent training. See you in the next chapter!

9
Deploying Deep RL Agents on Multiple Platforms

This chapter provides recipes to deploy your Deep RL agent models in applications targeting desktop, web, mobile, and beyond. The recipes serve as customizable templates that you can utilize to build and deploy your own Deep RL applications for your use cases. You will also learn how to export RL agent models for serving/deployment in various production-ready formats, such as **TensorFlow Lite**, **TensorFlow.js**, and **ONNX**, and learn how to leverage Nvidia **Triton** to launch production-ready RL-based AI services.

Specifically, the following recipes are covered in this chapter:

- Packaging Deep RL agents for mobile and IoT devices using TensorFlow Lite
- Deploying RL agents on mobile devices
- Packaging Deep RL agents for the web and Node.js using TensorFlow.js
- Deploying a Deep RL agent as a service
- Packaging Deep RL agents for cross-platform deployments

Technical requirements

The code in the book is extensively tested on Ubuntu 18.04 and Ubuntu 20.04 and should work with later versions of Ubuntu if Python 3.6+ is available. With Python 3.6+ installed along with the necessary Python packages, as listed before the start of each of the recipes, the code should run fine on Windows and Mac OSX too. It is advised to create and use a Python virtual environment named tf2rl-cookbook to install the packages and run the code in this book. Miniconda or Anaconda installation for Python virtual environment management is recommended.

The complete code for each recipe in each chapter will be available here: https://github.com/PacktPublishing/Tensorflow-2-Reinforcement-Learning-Cookbook.

Packaging Deep RL agents for mobile and IoT devices using TensorFlow Lite

This recipe will show how you can leverage the open source **TensorFlow Lite** (**TFLite**) framework for serving your Deep RL agents on mobile, IoT, and embedded devices. We will implement a complete script to build, train, and export an agent model that you can load into a mobile or embedded device. We will explore two methods to generate the TFLite model for our agent. The first method involves saving and exporting the agent models in TensorFlow's SavedModel file format and then using a command-line converter. The second method leverages the Python API to directly generate the TFLite models.

Let's get started!

Getting ready

To complete this recipe, you will first need to activate the tf2rl-cookbook Python/conda virtual environment. Make sure to update the environment to match the latest conda environment specification file (tfrl-cookbook.yml) in the cookbook's code repo. If the following imports work without issues, you are ready to get started:

```
import argparse
import os
import sys
from datetime import datetime

import gym
import numpy as np
```

```
import procgen  # Used to register procgen envs with Gym
registry
```

```
import tensorflow as tf
```

```
from tensorflow.keras.layers import Conv2D, Dense, Dropout,
Flatten, Input, MaxPool2D
```

Now, let's begin!

How to do it...

In the following steps, we will save space by focusing on the new and important pieces that are unique to this recipe. We will go through the model saving and export functionality and the different ways you can do that and keep the Actor, Critic, and Agent model definitions out of the following steps to save space. Please refer to the book's code repository for a complete implementation.

Let's get started:

1. First, it is important to set TensorFlow Keras's backend to use `float32` as the default representation for float values instead of the default `float64`:

```
tf.keras.backend.set_floatx("float32")
```

2. Next, let's create a handler for arguments passed to the script. We will also define a list of options for the training environments that can be chosen from for the `--env` flag:

```
parser = argparse.ArgumentParser(prog="TFRL-Cookbook-Ch9-
PPO-trainer-exporter-TFLite")
```

```
parser.add_argument(
    "--env", default="procgen:procgen-coinrun-v0",
    choices=["procgen:procgen-bigfish",
        "procgen:procgen-bossfight",
        "procgen:procgen-caveflyer",
        "procgen:procgen-chaser",
        "procgen:procgen-climber",
        "procgen:procgen-coinrun",
        "procgen:procgen-dodgeball",
        "procgen:procgen-fruitbot",
        "procgen:procgen-heist",
        "procgen:procgen-jumper",
```

```
            "procgen:procgen-leaper",
            "procgen:procgen-maze",
            "procgen:procgen-miner",
            "procgen:procgen-ninja",
            "procgen:procgen-plunder",
            "procgen:procgen-starpilot",
            "Pong-v4",
        ],
    )
```

3. We will add a few other arguments to ease the training and logging configuration of the agent:

```
parser.add_argument("--update-freq", type=int,
default=16)
parser.add_argument("--epochs", type=int, default=3)
parser.add_argument("--actor-lr", type=float,
default=1e-4)
parser.add_argument("--critic-lr", type=float,
default=1e-4)
parser.add_argument("--clip-ratio", type=float,
default=0.1)
parser.add_argument("--gae-lambda", type=float,
default=0.95)
parser.add_argument("--gamma", type=float, default=0.99)
parser.add_argument("--logdir", default="logs")


args = parser.parse_args()
```

4. Let's also set up logging so that we can visualize the agent's learning progress using TensorBoard:

```
logdir = os.path.join(
    args.logdir, parser.prog, args.env, \
    datetime.now().strftime("%Y%m%d-%H%M%S")
)
print(f"Saving training logs to:{logdir}")
writer = tf.summary.create_file_writer(logdir)
```

5. For the first export approach, we will define save methods for the Actor, Critic, and Agent classes in the following steps. We will start with the implementation of the save method in the Actor class to export the Actor model to TensorFlow's SavedModel format:

```
def save(self, model_dir: str, version: int = 1):
    actor_model_save_dir = os.path.join(
        model_dir, "actor", str(version), \
        "model.savedmodel"
    )
    self.model.save(actor_model_save_dir,
                    save_format="tf")
    print(f"Actor model saved at:\
            {actor_model_save_dir}")
```

6. Similarly, we will implement a save method for the Critic class to export the Critic model to TensorFlow's SavedModel format:

```python
def save(self, model_dir: str, version: int = 1):
    critic_model_save_dir = os.path.join(
        model_dir, "critic", str(version), \
        "model.savedmodel"
    )
    self.model.save(critic_model_save_dir,
                    save_format="tf")
    print(f"Critic model saved at:{
                        critic_model_save_dir}")
```

7. We can now add a save method for the Agent class that will utilize the Actor and Critic save method to save both the models needed by the Agent:

```python
def save(self, model_dir: str, version: int = 1):
    self.actor.save(model_dir, version)
    self.critic.save(model_dir, version)
```

8. Once the save() method is executed, it will generate two models (one for the Actor and one for the Critic) and save them in the specified directory on the filesystem with the directory structure and files similar to the one shown in the following figure:

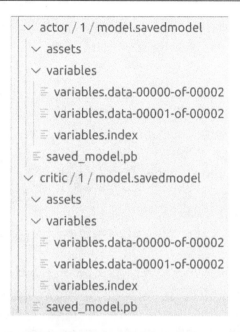

Figure 9.1 – TensorFlow SavedModel directory structure and file contents for the PPO RL agent

9. Once the SavedModel files are generated, we can use the tflite_convert command-line tool and specify the location of the Actor model's save directory. Refer to the following command for an example:

```
(tfrl-cookbook)praveen@desktop:~/tfrl-cookbook/
ch9$tflite_convert \
    --saved_model_dir=trained_models/ppo-procgen-coinrun/1/
actor/model.savedmodel \
    --output_file=trained_models/ppo-procgen-coinrun/1/
actor/model.tflite
```

10. Similarly, we can convert the Critic model using the following command:

```
(tfrl-cookbook)praveen@desktop:~/tfrl-cookbook/
ch9$tflite_convert \
    --saved_model_dir=trained_models/ppo-procgen-coinrun/1/
critic/model.savedmodel \
    --output_file=trained_models/ppo-procgen-coinrun/1/
critic/model.tflite
```

Hooray! We now have both the Actor and Critic models in TFLite format, which we can ship with our mobile applications. We will look at another approach that doesn't need us to (manually) switch to the command line to convert the model.

11. There's another approach to export the Agent model to the TFLite format. We will be implementing it in the following steps, starting with the `save_tflite` method for the `Actor` class:

```python
def save_tflite(self, model_dir: str, version: int =\
    1):
    """Save/Export Actor model in TensorFlow Lite
    format"""
    actor_model_save_dir = os.path.join(model_dir,\
                        "actor", str(version))
    model_converter = \
        tf.lite.TFLiteConverter.from_keras_model(
                                self.model)
    # Convert model to TFLite Flatbuffer
    tflite_model = model_converter.convert()
    # Save the model to disk/persistent-storage
    if not os.path.exists(actor_model_save_dir):
        os.makedirs(actor_model_save_dir)
    actor_model_file_name = os.path.join(
            actor_model_save_dir, "model.tflite")
    with open(actor_model_file_name, "wb") as \
    model_file:
        model_file.write(tflite_model)
    print(f"Actor model saved in TFLite format at:\
        {actor_model_file_name}")
```

12. Similarly, we will implement the `save_tflite` method for the `Critic` class:

```python
def save_tflite(self, model_dir: str, version: \
    int = 1):
    """Save/Export Critic model in TensorFlow Lite
    format"""
    critic_model_save_dir = os.path.join(model_dir,
                        "critic", str(version))
    model_converter = \
        tf.lite.TFLiteConverter.from_keras_model(
                                self.model)
```

```
# Convert model to TFLite Flatbuffer
tflite_model = model_converter.convert()
# Save the model to disk/persistent-storage
if not os.path.exists(critic_model_save_dir):
    os.makedirs(critic_model_save_dir)
critic_model_file_name = os.path.join(
        critic_model_save_dir, "model.tflite")
with open(critic_model_file_name, "wb") as \
model_file:
    model_file.write(tflite_model)
print(f"Critic model saved in TFLite format at:\
    {critic_model_file_name}")
```

13. The Agent's class can then call the `save_tflite` method on the Actor and Critic using its own `save_tflite` method, as shown in the following code snippet:

```
def save_tflite(self, model_dir: str, version: \
int = 1):
    # Make sure `toco_from_protos binary` is on
    # system's PATH to avoid TFLite ConverterError
    toco_bin_dir = os.path.dirname(sys.executable)
    if not toco_bin_dir in os.environ["PATH"]:
        os.environ["PATH"] += os.pathsep + \
                              toco_bin_dir
    print(f"Saving Agent model (TFLite) to:{
                            model_dir}\n")
    self.actor.save_tflite(model_dir, version)
    self.critic.save_tflite(model_dir, version)
```

Notice that we added the `bin` directory of the current (`tfrl-cookbook`) Python environment to the system's PATH environment variable to make sure the `toco_from_protos` binary is found when the TFLite converter invokes the model conversion.

14. To sum up, we can finalize the `main` function to instantiate the agent and train and save the model in TFLite model file format:

```python
if __name__ == "__main__":
    env_name = args.env
    env = gym.make(env_name)
    agent = PPOAgent(env)
    agent.train(max_episodes=1)
    # Model saving
    model_dir = "trained_models"
    agent_name = f"PPO_{env_name}"
    agent_version = 1
    agent_model_path = os.path.join(model_dir, \
                                    agent_name)
    agent.save_tflite(agent_model_path, agent_version)
```

That completes our recipe. Let's recap with some important details to understand the recipe better.

How it works...

We first set TensorFlow Keras's backend to use `float32` as the default representation for float values. This is because, otherwise, TensorFlow would use the default `float64` representation, which is not supported by TFLite (for performance reasons) as it is targeted towards running on embedded and mobile devices.

Then, we defined a list of choices for the `--env` argument. This is important to make sure that the environment's observation and action spaces are compatible with the agent's model. In this recipe, we used a PPO agent with Actor and Critic networks that expect image observations and produce actions in discrete space. You can swap the agent code with the PPO implementations from one of the earlier chapters that use different state/observation spaces and action spaces. You could also replace the agent with a different agent algorithm altogether. You will find a bonus recipe that exports a DDPG agent TFLite model in the book's code repository for this chapter.

We discussed two approaches to save and convert our Agent models to TFLite format. The first approach allowed us to generate a TensorFlow SavedModel file format first and then convert it to the TFLite model file format using the `tflite_convert` command-line tool. In the second approach, we used TFLite's Python API to directly (in-memory) convert and save the agent's models in TFLite (Flatbuffer) format. We made use of the `TFLiteConverter` module, which ships with the official TensorFlow 2.x Python package. A summary of different ways to export the RL agent's model using the API is provided in the following figure:

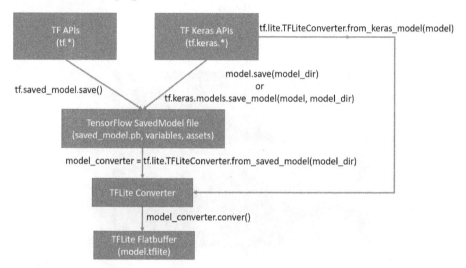

Figure 9.2 – Converting TensorFlow 2.x models to TensorFlow Lite Flatbuffer format

You can learn more about the TFLite model format here: `https://www.tensorflow.org/lite`.

It's time to hop on to the next recipe!

Deploying RL agents on mobile devices

Mobile is the most-targeted platform due to its high customer reach compared to other platforms. The global mobile application market size is projected to reach USD 407.32 billion by 2026 according to `https://www.alliedmarketresearch.com/mobile-application-market`. Such a huge market opens several opportunities for infusing RL-based Artificial Intelligence. Android and iOS are the two main OS platforms in this space. While IOS is a popular platform, building apps for iOS requires a Mac to develop the apps. We will therefore develop an Android app using the Android SDK, which is more widely accessible. If you are an iOS app developer, you may be able to adapt parts of this recipe to your app.

This recipe provides ways for you to deploy trained RL agent models on mobile and/or IoT devices using the TensorFLow Lite framework. You will also have access to a sample RL Table Tennis Android app that you can use as a testbed to deploy your RL agent or develop your own ideas and apps:

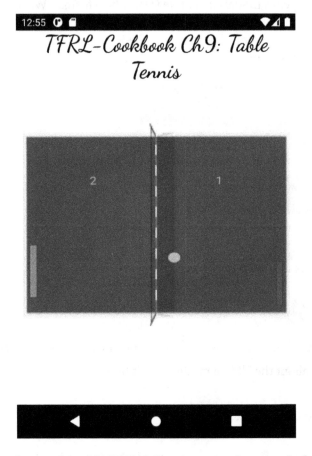

Figure 9.3 – A screenshot of the RL Table Tennis app running on an Android device

Let's get started!

Getting ready

We will be using Android Studio to set up and develop the sample RL Android app. Download and install Android Studio from the official website: `https://developer.android.com/studio`. Using the default install location is recommended. Once installed, run Android Studio to start the **Android Studio Setup Wizard**. Follow through the setup process and make sure the latest Android SDK, Android SDK command-line tools, and the Android SDK build tools are marked for installation.

To run the application once complete, you have two options: 1. Run it on your Android phone 2. Run it in the Android virtual device emulator. Follow the setup instructions depending on your choice:

- Running it on your Android phone:

  a) Enable developer options and USB debugging in Android settings. Detailed instructions are available here: `https://developer.android.com/studio/debug/dev-options`.

  b) If you are on Windows, install the Google USB driver: `https://developer.android.com/studio/run/win-usb`.

  c) Connect your phone to your computer using a USB cable and, if prompted, allow your computer to access your phone.

  d) Run `adb devices` to make sure your phone is detected. If your phone is not detected, make sure the drivers are installed and ADB debugging is enabled on your phone. You can follow the Android official guide here for detailed instructions: `https://developer.android.com/studio/run/device#setting-up`.

- Running it in the Android emulator:

  a) Launch Android Studio, click on the **AVD Manager** icon and select **Create Virtual Device**.

  b) Choose a device and select **Next**.

  c) Choose an x86 or x86_64 image for the Android version you want to emulate and complete the process.

  d) Click **Run** in the AVD Manager toolbar to launch the emulator.

Once you have the device set up, navigate to the code directory for this recipe under the `src/ch9-cross-platform-deployment` directory. You will find a sample Android application with a directory structure and contents like the one shown in the following screenshot:

> .gradle

> .idea

> app

> gradle

◆ .gitignore

🐘 build.gradle

Ⓙ gradle.properties

≡ gradlew

🪟 gradlew.bat

Ⓙ local.properties

🐘 settings.gradle

Figure 9.4 – Directory structure and contents of the sample Android app

Once you have the sample code base to work with, move on to the next section to see how to prepare our RL agent model and build the app.

How to do it...

We'll start with the RL agent model preparation and then build a simple, two-player Table Tennis app where you can play against the agent. Follow the steps listed here:

1. Export your RL agent's (Actor) model to TFLite format using the previous recipe discussed in this chapter. For example, you can run the previous recipe to train a PPO agent for the Pong-v4 environment and use the generated model.tflite file in the trained_models/actor/1/ directory. Place the model in the Android app's app/src/assets/ directory as highlighted in the figure here:

Figure 9.5 – RL agent model.tflite location in Android app src

2. Edit the app's `dependencies` section in the `build.gradle` file to include the `tensorflow-lite` dependency:

```
dependencies {
    implementation fileTree(dir: 'libs', include: \
                            ['*.jar'])
    implementation 'org.tensorflow:tensorflow-lite:+'
}
```

3. Add a member method to load the `agent/model.tflite` from the `assets` folder and return a `MappedByteBuffer`:

```
MappedByteBuffer loadModelFile(AssetManager \
    assetManager) throws IOException {
AssetFileDescriptor fileDescriptor = \
    assetManager.openFd("agent/model.tflite");
FileInputStream inputStream = new \
    FileInputStream(
        fileDescriptor.getFileDescriptor());
FileChannel fileChannel = \
            inputStream.getChannel();
long startOffset = \
    fileDescriptor.getStartOffset();
long declaredLength = \
    fileDescriptor.getDeclaredLength();
return fileChannel.map(
    FileChannel.MapMode.READ_ONLY, \
    startOffset, declaredLength);
}
```

4. We can now create a new TFLite interpreter like so:

```
interpreter = new Interpreter(loadModelFile(assetManager),
                           new Interpreter.Options());
```

5. The interpreter is ready. Let's prepare the input. First, let's define some constants based on what we know from our agent training:

```
static final int BATCH_SIZE = 1;
 static final int OBS_IMG_WIDTH = 160;
 static final int OBS_IMG_HEIGHT = 210;
 static final int OBS_IMG_CHANNELS = 3;
 // Image observation normalization
 static final int IMAGE_MEAN = 128;
 static final float IMAGE_STD = 128.0f;
```

6. Let's now implement a method to convert image data in `BitMap` format to `ByteArray`:

```
ByteBuffer convertBitmapToByteBuffer(Bitmap bitmap) {
    ByteBuffer byteBuffer;
    byteBuffer = ByteBuffer.allocateDirect(4 * \
                BATCH_SIZE * OBS_IMG_WIDTH * \
                OBS_IMG_HEIGHT * OBS_IMG_CHANNELS);
    byteBuffer.order(ByteOrder.nativeOrder());
    int[] intValues = new int[OBS_IMG_WIDTH * \
                            OBS_IMG_HEIGHT];
    bitmap.getPixels(intValues,0, bitmap.getWidth(),\
        0, 0, bitmap.getWidth(), bitmap.getHeight());
    int pixel = 0;
    for (int i = 0; i < OBS_IMG_HEIGHT; ++i) {
        for (int j = 0; j < OBS_IMG_WIDTH; ++j) {
            final int val = intValues[pixel++];

                byteBuffer.putFloat((((val >> 16) &\
                0xFF)-IMAGE_MEAN)/IMAGE_STD);
                byteBuffer.putFloat((((val >> 8) & \
                0xFF)-IMAGE_MEAN)/IMAGE_STD);
```

```
                              byteBuffer.putFloat((((val) & 0xFF)-\
                              IMAGE_MEAN)/IMAGE_STD);
              }
          }
          return byteBuffer;
      }
```

7. We can now run the image observations from the Table Tennis game through the Agent model to get the action:

```
ByteBuffer byteBuffer = convertBitmapToByteBuffer(bitmap);
int[] action = new int[ACTION_DIM];
interpreter.run(byteBuffer, action);
```

Those are all the main ingredients for this recipe! You can run them in a loop to generate actions per observation/game frame or customize them however you like! Let's look at how to run the app on an Android device using Android Studio in the following steps.

8. Launch Android Studio. You will see a screen like this:

Figure 9.6 – Android Studio welcome screen

Let's proceed to the next step.

9. Click on the **Open an Existing Project** option and you will see a popup asking you to choose the directory on your filesystem. Navigate to the folder where you have cloned the book's code repo or your fork, and browse to this recipe's folder under Chapter 9 as shown in the figure here:

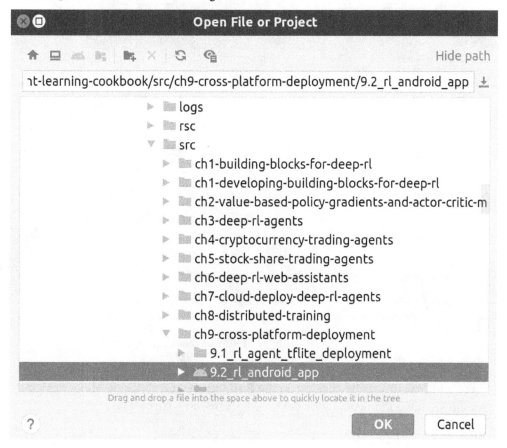

Figure 9.7 – File/project picker interface to choose the RL Android app

You will notice that Android Studio has already identified our app and shows the directory with an Android symbol.

10. Once you click **OK**, Android Studio will open with the app's code and will look like the following figure:

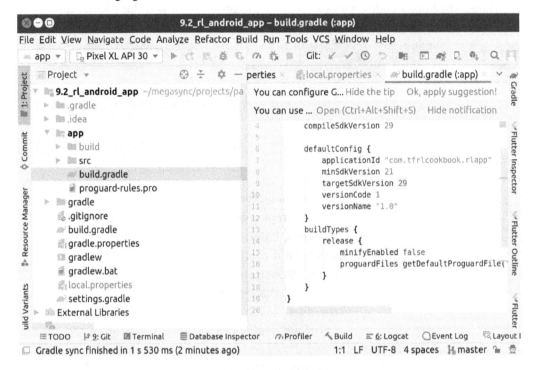

Figure 9.8 – Android Studio with the TFRL-Cookbook's RL app loaded

So far, so good!

11. Let's build the project by clicking on the **Build** menu and choosing **Make Project**, as shown in the following figure (or by simply pressing *Ctrl + F9*):

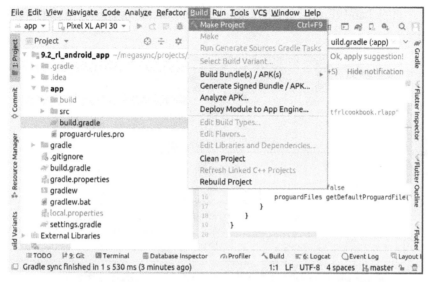

Figure 9.9 – Building the RL Android app using the Make Project option

This process may take some time to complete and you may see useful status messages in the **Build** information tab.

12. Once the build process completes, you will see **BUILD SUCCESSFUL** in the **Build** console output, as shown in the following figure:

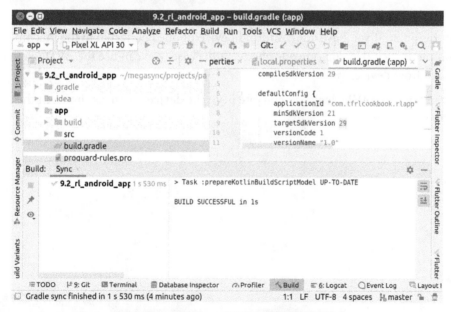

Figure 9.10 – RL Android app BUILD SUCCESSFUL message

The build process generates a `.apk` file, which can be run on an Android device.

13. Let's go ahead and run the app by using the **Run** menu, as shown in the following figure:

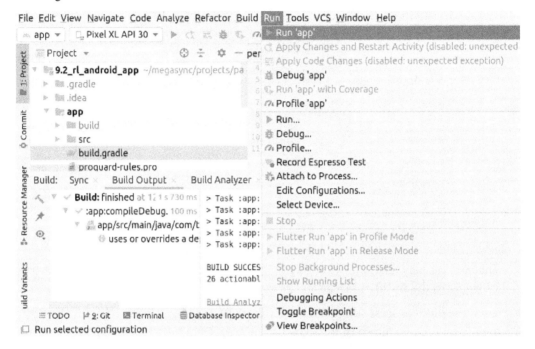

Figure 9.11 – The Run menu option to run the RL app in Android Studio

At this point, if you have your Android device/phone connected to the machine, you can launch that app on your phone. Otherwise, you can use the AVD to emulate an Android device.

14. Let's choose an AVD device to emulate from the device menu, as shown in the following figure:

Figure 9.12 – Choose the AVD to emulate an Android device

We are now ready with the device to run the app.

15. Let's go ahead and launch/run the app! You can use the **Run 'app'** button from the **Run** menu, as shown in the following figure:

Figure 9.13 – Run 'app' command to launch the app

That should launch the app on the AVD emulator (or on your phone if you chose it).

16. The app should launch on the Android device and should look something like the following figure:

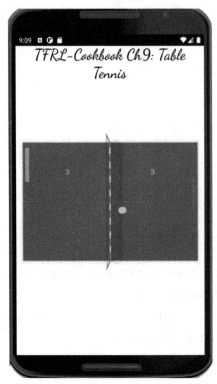

Figure 9.14 – The TFRL-Cookbook RL app running on an Android (emulated) device

Congratulations!

That completes our recipe. Head to the next section to learn more about the recipe.

How it works...

In the previous recipe, we saw how you can export your Deep RL agent's model to the TFLite format. The previous recipe generated two `model.tflite` files: one for the Actor and another for the Critic.

> **Note**
>
> You can train any agent algorithm of your choice following the recipes previously discussed in this book and use the recipe titled *Packaging Deep RL agents for mobile and IoT devices using TensorFlow Lite* in this chapter to obtain the Actor `model.tflite` file used in this recipe.

As you may recall from *Chapter 3, Implementing Advanced Deep RL Algorithms on Deep RL Agents*, the Actor component is responsible for generating actions according to the learned policy and the Critic component estimates the state or state-action value. When it comes to deploying RL agents, we are more interested in the action generated by the agent than the predicted state or state-action values. Therefore, we only used the agent's Actor model for deployment purposes in this recipe.

We first included the TFLite dependency by updating the app's `gradle.build` file. We then added a method named `loadModelFile` to load the agent's model (`model.tflite`). This returns a `MappedByteBuffer` object, which is needed to initialize a TFLite interpreter instance. Once the agent's model is loaded and a TFLite interpreter instance is created, we can run the interpreter with valid inputs to get the agent's actions. In order to make sure the inputs are in a valid format, we converted the image data from `BitMap` format to `ByteBuffer` format. We also defined the image observation width, height, number of channels, and so on based on the observation space of the environment we used to train the RL agent.

The action returned by the agent's model in *Step 7* can be used to actuate/move, say, the red paddle in the Table Tennis game and repeat the preceding steps for each new observation in a loop to make the agent play against itself or a human!

We then saw how to launch the app using Android Studio and then concluded the recipe. Hope you had fun!

Let's march on to the next recipe whenever you are ready.

Packaging Deep RL agents for the web and Node.js using TensorFlow.js

JavaScript is the language of choice when it comes to developing web applications due to its versatility both as a frontend as well as a backend programming language that can be executed by a web browser or using Node.js. The ability to run out RL agents on the web will unlock several new pathways for deploying RL agents in web apps. This recipe will show how you can train and export RL agent models into a format that you can then use in your JavaScript applications that can be run directly in the browser or in a Node.js environment. The TensorFlow.js (TF.js) library allows us to use JavaScript to run existing models or even train/retrain new models. We will use the `tensorflowjs` Python module to export our agent's model to a supported format that can be imported into JavaScript-based web or desktop (Node.js/Electron) apps. We will explore two approaches to export the Agent model to the TF.js layers format.

Let's get started!

Getting ready

To complete this recipe, you will first need to activate the `tf2rl-cookbook` Python/ conda virtual environment. Make sure to update the environment to match the latest conda environment specification file (`tfrl-cookbook.yml`) in the cookbook's code repo. If the following imports work without issues, you are ready to get started:

```
import argparse
import copy
import os
import random
from collections import deque
from datetime import datetime

import gym
import numpy as np
import tensorflow as tf
import tensorflowjs as tfjs
from tensorflow.keras.layers import (
    Conv2D,
    Dense,
    Dropout,
    Flatten,
    Input,
    Lambda,
    MaxPool2D,
)

import webgym
```

Now, let's begin!

How to do it...

In the following text, we will save space by focusing on the new and important pieces that are unique to this recipe. We will go through the model saving and export functionality and the different ways you can do that and keep the Actor, Critic, and Agent model definitions out of the following steps to save space. Please refer to the book's code repository for a complete implementation, including the training and logging methods.

Let's get started:

1. Let's first set up a command-line argument parser to allow easy customization of the script:

```
parser = argparse.ArgumentParser(
    prog="TFRL-Cookbook-Ch9-DDPGAgent-TensorFlow.
js-exporter"
)
parser.add_argument("--env",
default="MiniWoBSocialMediaMuteUserVisualEnv-v0")
parser.add_argument("--actor_lr", type=float,
default=0.0005)
parser.add_argument("--critic_lr", type=float,
default=0.001)
parser.add_argument("--batch_size", type=int, default=64)
parser.add_argument("--tau", type=float, default=0.05)
parser.add_argument("--gamma", type=float, default=0.99)
parser.add_argument("--train_start", type=int,
                    default=2000)
parser.add_argument("--logdir", default="logs")

args = parser.parse_args()
```

2. Let's also set up logging so that we can visualize the agent's learning progress using TensorBoard:

```
logdir = os.path.join(
    args.logdir, parser.prog, args.env, \
    datetime.now().strftime("%Y%m%d-%H%M%S")
)
print(f"Saving training logs to:{logdir}")
writer = tf.summary.create_file_writer(logdir)
```

3. For the first export approach, we will define `save_h5` methods for the `Actor`, `Critic`, and `Agent` classes in the following steps. We will start with the implementation of the `save_h5` method in the `Actor` class to export the Actor model to Keras's h5 format:

```python
def save_h5(self, model_dir: str, version: int = 1):
    actor_model_save_dir = os.path.join(
        model_dir, "actor", str(version), "model.h5"
    )
    self.model.save(actor_model_save_dir, \
                    save_format="h5")
    print(f"Actor model saved at:\
        {actor_model_save_dir}")
```

4. Similarly, we will implement a `save` method for the `Critic` class to export the Critic model to Keras's h5 format:

```python
def save_h5(self, model_dir: str, version: int = 1):
    critic_model_save_dir = os.path.join(
        model_dir, "critic", str(version), "model.h5"
    )
    self.model.save(critic_model_save_dir, \
                    save_format="h5")
    print(f"Critic model saved at:\
        {critic_model_save_dir}")
```

5. We can now add a `save` method for the `Agent` class that will utilize the Actor and Critic `save` method to save both the models needed by the agent:

```python
def save_h5(self, model_dir: str, version: int = 1):
    self.actor.save_h5(model_dir, version)
    self.critic.save_h5(model_dir, version)
```

6. Once the `save_h5 ()` method is executed, the `save` method will generate two models (one for the Actor and one for the Critic) and save them in the specified directory on the filesystem with a directory structure and files like the one shown in the following figure:

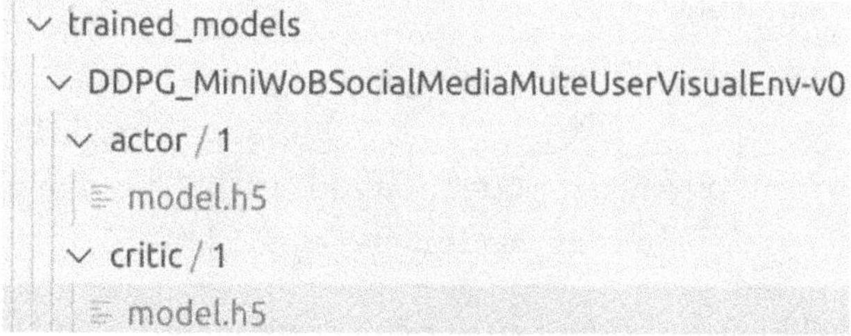

Figure 9.15 – Directory structure and file contents for the DDPG RL agent with the save_h5 model export

7. Once the .h5 files are generated, we can use the tensorflowjs_converter command-line tool and specify the location of the Actor model's save directory. Refer to the following command for an example:

```
(tfrl-cookbook)praveen@desktop:~/tfrl-cookbook/
ch9$tensorflowjs_converter --input_format keras \
                    actor/1/model.h5 \
                    actor/t1/model.tfjs
```

8. Similarly, we can convert the Critic model using the following command:

```
(tfrl-cookbook)praveen@desktop:~/tfrl-cookbook/
ch9$tensorflowjs_converter --input_format keras \
                    critic/1/model.h5 \
                    critic/t1/model.tfjs
```

Hooray! We now have both the Actor and Critic models in the TF.js layers format. We will look at another approach that doesn't need us to (manually) switch to the command line to convert the model.

9. There's another approach to export the Agent model to the TF.js layers format. We will be implementing it in the following steps, starting with the save_tfjs method for the Actor class:

```
def save_tfjs(self, model_dir: str, version: \
int = 1):
    """Save/Export Actor model in TensorFlow.js
    supported format"""
    actor_model_save_dir = os.path.join(
        model_dir, "actor", str(version), \
        "model.tfjs"
```

```
    )
    tfjs.converters.save_keras_model(self.model,\
                            actor_model_save_dir)
    print(f"Actor model saved in TF.js format at:\
            {actor_model_save_dir}")
```

10. Similarly, we will implement the save_tfjs method for the Critic class:

```
def save_tfjs(self, model_dir: str, version: \
int = 1):
    """Save/Export Critic model in TensorFlow.js
    supported format"""
    critic_model_save_dir = os.path.join(
        model_dir, "critic", str(version), \
        "model.tfjs"
    )
    tfjs.converters.save_keras_model(self.model,\
                            critic_model_save_dir)
    print(f"Critic model saved TF.js format \
            at:{critic_model_save_dir}")
```

11. The Agent class can then call the save_tfjs method on the Actor and Critic using its own save_tfjs method, as shown in the following code snippet:

```
def save_tfjs(self, model_dir: str, version: \
int = 1):
    print(f"Saving Agent model to:{model_dir}\n")
    self.actor.save_tfjs(model_dir, version)
    self.critic.save_tfjs(model_dir, version)
```

12. When the Agent's save_tfjs method gets executed, the Actor and Critic models in the TF.js layers format will be generated and will have a directory structure and file contents like the one shown in the following figure:

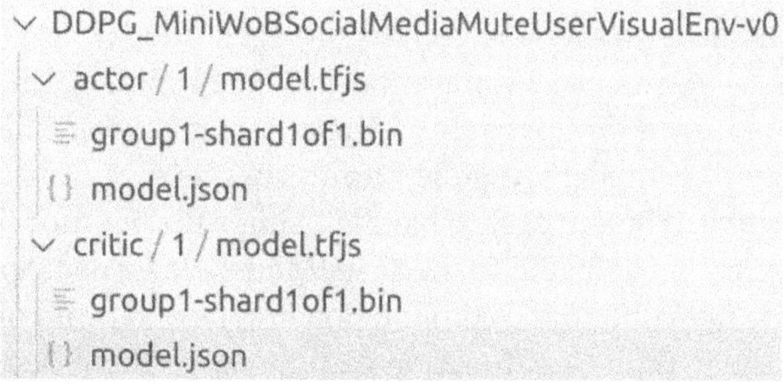

DDPG_MiniWoBSocialMediaMuteUserVisualEnv-v0
- actor / 1 / model.tfjs
  - group1-shard1of1.bin
  - {} model.json
- critic / 1 / model.tfjs
  - group1-shard1of1.bin
  - {} model.json

Figure 9.16 – Directory structure and file contents for the DDPG RL agent with the save_tfjs model export

13. To sum up, we can finalize the `main` function to instantiate the agent and train and save the model in the TF.js layers format directly using the Python API:

```python
if __name__ == "__main__":
    env_name = args.env
    env = gym.make(env_name)
    agent = PPOAgent(env)
    agent.train(max_episodes=1)
    # Model saving
    model_dir = "trained_models"
    agent_name = f"PPO_{env_name}"
    agent_version = 1
    agent_model_path = os.path.join(model_dir, \
                                    agent_name)
    # agent.save_h5(agent_model_path, agent_version)
    agent.save_tfjs(agent_model_path, agent_version)
```

14. You can now take the TF.js model and deploy it in your web app, Node.js app, Electron app, or any other JavaScript/TypeScript-based applications. Let's recap some of the key items we used in this recipe in the next section.

How it works...

In this recipe, we used a DDPG agent with Actor and Critic networks that expect image observations and produce actions in continuous space. You can swap the agent code with the DDPG implementations from one of the earlier chapters that use different state/observation spaces and action spaces. You could also replace the agent with a different agent algorithm altogether. You will find a bonus recipe that exports a PPO agent TF.js model in the book's code repository for this chapter.

We discussed two approaches to save and convert our agent models to TF.js format. The first approach allowed us to generate a Keras model in H5 format, which is a short form of HDF5, which is an acronym for Hierarchical Data Format version 5 file format. We then converted it to the TF.js model using the `tensorflowjs_converter` command-line tool. While it is lightweight and easy to handle a single file per model, the Keras HDF5 model has limitations compared to the SavedModel file format. Specifically, the Keras HDF5 models do not contain the computation graphs of custom objects/layers and therefore will require the Python class/function definitions for these custom objects to reconstruct the model during runtime. Moreover, in the cases when we add loss terms and metrics outside the model class definition (using `model.add_loss()` or `model.add_metric()`), these are not exported in the HDF5 model file.

In the second approach, we used the `tensorflowjs` Python module to directly (in memory) convert and save the agent's models in the TF.js layers format.

You can learn more about TF.js here: `https://www.tensorflow.org/js`.

It's time for the next recipe!

Deploying a Deep RL agent as a service

Once you train your RL agent to solve a problem or business need, you will want to deploy it as a service – more likely than offering the trained agent model as a product due to several reasons, including scalability and model-staleness limitations. You will want to have a way to update the agent model with new versions and you will not want to maintain or offer support for multiple versions or older versions of your agent if you sell it as a product. You will need a solid and well-tested mechanism to offer your RL agent as an AI service that allows customizable runtimes (different frameworks, and CPU/GPU support), easy model upgrades, logging, performance monitoring, and so on.

To serve all such needs, we will be using NVIDIA's Triton server as the backend for serving our agent as a service. Triton serves as a unifying inference framework for the deployment of AI models at scale in production. It supports a wide variety of deep learning frameworks including TensorFlow2, PyTorch, ONNX, Caffe2, and others, including custom frameworks, and offers several other production-quality features and optimizations, such as concurrent model execution, dynamic batching, logging, and performance and health monitoring.

Let's get started with our recipe!

Getting ready

To complete this recipe, you will first need to activate the `tf2rl-cookbook` Python/conda virtual environment. Make sure to update the environment to match the latest conda environment specification file (`tfrl-cookbook.yml`) in the cookbook's code repo. You will also need to make sure you have the latest NVIDIA GPU drivers installed on your machine that supports the GPU you have. You will also need Docker set up on your machine. If you haven't installed Docker, you can follow the official instructions here to set up Docker for your OS: `https://docs.docker.com/get-started/`.

Now, let's begin!

How to do it...

In the following text, we will save space by focusing on the service we will be building. We will keep the contents of the agent training scripts out of the text, but you will find the scripts in the book's code repository under `ch9-cross-platform-deployment`.

Let's get started:

1. First, you will want to train, save, and export the agent that you want to host as a service. You can use the sample `agent_trainer_saver.py` script to train a PPO agent for one of the tasks in the Webgym suite of environments using the following command:

    ```
    $ python agent_trainer_saver.py
    ```

    Once the trained agent model is ready, we can move on to the next step.

    Check the NVIDIA framework support matrix at `https://docs.nvidia.com/deeplearning/frameworks/support-matrix/index.html` to find the container image version (in yy.mm format) that supports your NVIDIA GPU driver version. For example, if you have installed NVIDIA driver version 450.83 (find out by running `nvidia-smi`), then container versions built with CUDA 11.0.3 or lower, such as container version 20.09 or older, will work.

2. Once you have identified the suitable container version, say yy.mm, you can use Docker to pull the NVIDIA Triton server image using the following command:

    ```
    praveen@desktop:~$ docker pull nvcr.io/nvidia/
    tritonserver:yy.mm-py3
    ```

3. Change the yy.mm placeholder to the version you have identified. For example, to pull the container version 20.09, you would run the following command:

```
praveen@desktop:~$ docker pull nvcr.io/nvidia/
tritonserver:20.09-py3
```

4. When you run the agent_trainer_saver script, the trained models are stored in the trained_models directory with the following directory structure and contents:

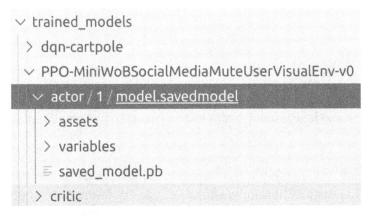

Figure 9.17 – Directory structure and contents of the exported trained models

The trained_models/actor directory will be the root directory for our model repository store when serving with Triton.

5. We are now ready to serve our agent's actions as a service! To launch the service, run the following command:

```
$ docker run --shm-size=1g --ulimit memlock=-1 --ulimit
stack=67108864 --gpus=1 --rm -p8000:8000 -p8001:8001
-p8002:8002 -v/full/path/to/trained_models/actor:/
models nvcr.io/nvidia/tritonserver:yy.mm-py3 tritonserver
--model-repository=/models --strict-model-config=false
--log-verbose=1
```

Remember to update the Docker volume path after the -v flag to point to the trained_models/actor folder on your serving machine. Also remember to update the yy.mm value to reflect your container version (20.3, for example).

6. If you want to serve the agent model from a machine that does not have a GPU (not recommended), you can simply omit the –gpus=1 flag to instruct the Triton server to serve using CPUs only. The command will look like this:

```
$ docker run   --shm-size=1g --ulimit memlock=-1 --ulimit
stack=67108864 --rm -p8000:8000 -p8001:8001 -p8002:8002
-v/full/path/to/trained_models/actor:/models nvcr.io/
nvidia/tritonserver:yy.mm-py3 tritonserver --model-
repository=/models --strict-model-config=false
--log-verbose=1
```

7. If you run into issues serving your agent models, check the trained_models/ actor/config.pbtxt file, which describes the model configuration. While Triton can automatically generate the config.pbtxt file from TensorFlow SavedModels, it may not work well for all, especially a custom policy network implementation. If you are using the agent_trainer_saver script to export a trained PPO agent, you can use the following config.pbtxt. We will discuss the model config in the next few steps:

```
{
    "name": "actor",
    "platform": "tensorflow_savedmodel",
    "backend": "tensorflow",
    "version_policy": {
        "latest": {
            "num_versions": 1
        }
    },
    "max_batch_size": 1,
```

8. We will continue to specify the input (state/observation) space/dimension configuration:

```
    "input": [
        {
            "name": "input_1",
            "data_type": "TYPE_FP64",
            "dims": [
                64,
                64,
```

```
                3
            ],
            "format": "FORMAT_NHWC"
        }
    ],
```

9. Next, we will specify the output (action space):

```
    "output": [
        {
            "name": "lambda",
            "data_type": "TYPE_FP64",
            "dims": [
                2
            ]
        },
        {
            "name": "lambda_1",
            "data_type": "TYPE_FP64",
            "dims": [
                2
            ]
        }
    ],
```

10. Let's also specify the instance group, optimization parameters, and so on:

```
    "batch_input": [],
    "batch_output": [],
    "optimization": {
        "priority": "PRIORITY_DEFAULT",
        "input_pinned_memory": {
            "enable": true
        },
        "output_pinned_memory": {
            "enable": true
        }
```

```
        },
        "instance_group": [
            {
                "name": "actor",
                "kind": "KIND_CPU",
                "count": 1,
                "gpus": [],
                "profile": []
            }
        ],
```

11. The final set of config parameters required for the `config.pbtxt` file is listed here:

```
        "default_model_filename": "model.savedmodel",
        "cc_model_filenames": {},
        "metric_tags": {},
        "parameters": {},
        "model_warmup": []
    }
```

12. Hooray! Our agent as a service is live. At this point, you can run the same commands we discussed above on a cloud/remote server/VPS if you would like to offer this service on the public web/a network. Let's quickly send a query to the server to make sure everything went as expected:

```
$curl -v localhost:8000/v2/health/ready
```

13. If the agent model is being served without issues, you will see an output similar to the following:

```
...
< HTTP/1.1 200 OK
< Content-Length: 0
< Content-Type: text/plain
```

14. You can also use a full-fledged sample client app to query the agent service to get the prescribed action. Let's quickly set up the tools and libraries we need for running a Triton client. You can use Python pip to install the dependencies, as shown in the following command snippet:

```
$ pip install nvidia-pyindex
$ pip install tritonclient[all]
```

15. Optionally, to be able to run the performance analyzer (`perf_analyzer`), you will need to install the libb64-dev system package using the following command:

```
$ sudo apt update && apt install libb64-dev
```

16. You now have all the dependencies to run the sample Triton client app:

```
$ python sample_client_app.py
```

That completes our recipe! Let's look into some of the details of what we accomplished in this recipe in the next section.

How it works...

Our recipe had three sections:

1. Train, save, and export;

2. Deploy;

3. Launch client.

The first section covered the agent training, saving, and exporting routine. In this section, we first picked the RL environment and agent algorithm we wanted to train. We then utilized one of the many training strategies we discussed earlier in this book to train the agent model. We then used the model saving and export methods we discussed in the previous recipes of this chapter to export the trained agent model in TensorFlow's SavedModel file format. As you may recall, we followed a specific directory structure and file naming convention when we saved and exported our agent model. This convention aligns with the model repository conventions used by the NVIDIA Triton server and thus allows the models we export to be easily served with the production-ready Triton server. Moreover, the organization allows us to manage multiple versions of the agent model concurrently easily.

In the second section, we saw how we can deploy the exported agent model using NVIDIA's Triton server. You can learn more about NVIDIA's Triton here: `https://developer.nvidia.com/nvidia-triton-inference-server`.

We saw how easy it is to serve our agent using a production-grade serving backend. We can easily run the Docker container on a remote/cloud server or VPS to deploy this service out on the web.

Finally, once the service was launched, we saw how a client can avail of the service by sending action requests with appropriate input/observation data from a test environment.

That's it for this recipe! Let's move on to the final recipe of this chapter to wrap things up.

Packaging Deep RL agents for cross-platform deployment

Although the grandest success of Deep RL has been in the domain of game playing (Atari, Chess, Go, Shogi) and simulated robotics, real-world applications are starting to emerge where Deep RL agents show a lot of promise and value. Deploying Deep RL agents to a variety of physical form factors such as embedded controllers, computers, autonomous cars, drones, and other robots, and so on is expected soon. Differences in hardware processors (CPU, GPU, TPU, FPGA, ASIC), operating systems (Linux, Windows, OSX, Android), architectures (x86, ARM), and form factors (server, desktop, mobile, IoT, embedded systems, and so on) make the deployment process challenging. This recipe includes guidelines around how you can leverage the TensorFlow 2.x framework's ecosystem of libraries, tools, and utilities to package Deep RL agent models suitable for deployments to the web, mobile, IoT, embedded systems, robots, and desktop platforms.

This recipe provides a complete script to build, train, and package a Deep RL agent in multiple formats that can be used to deploy/serve using TensorFlow Serving, TensorFlow Hub, TensorFlow.js, TensorFlow Lite, NVIDIA Triton, ONNX, ONNX.js, Clipper, and most other serving frameworks built for deep learning models.

Let's get started!

Getting ready

To complete this recipe, you will first need to activate the `tf2rl-cookbook` Python/conda virtual environment. Make sure to update the environment to match the latest conda environment specification file (`tfrl-cookbook.yml`) in the cookbook's code repo. If the following imports work without issues, you are ready to get started:

```
import argparse
import os
import sys
from datetime import datetime


import gym
import keras2onnx
import numpy as np
import procgen  # Used to register procgen envs with Gym
registry
import tensorflow as tf
import tensorflowjs as tfjs
from tensorflow.keras.layers import Conv2D, Dense, Dropout,
Flatten, Input, MaxPool2D
```

Now, let's begin!

How to do it...

In the following text, we will save space by focusing on the new and important pieces that are unique to this recipe. We will focus on the various model saving and export functionalities and keep the Actor, Critic, and Agent model definitions out of the following steps to save space. Please refer to the book's code repository for a complete implementation. We will start implementing the model's save/export methods one after the other for the Actor first and then repeat the steps for the Critic in the subsequent steps, and finally complete the agent implementation.

Let's get started:

1. First, it is important to set TensorFlow Keras's backend to use `float32` as the default representation for float values instead of the default `float64`:

    ```
    tf.keras.backend.set_floatx("float32")
    ```

2. We will begin with the various save/export method implementations for the Actor in the following few steps. Let's implement the `save` method to save and export the Actor model to TensorFlow's `SavedModel` format:

```python
def save(self, model_dir: str, version: int = 1):
    actor_model_save_dir = os.path.join(
        model_dir, "actor", str(version), \
        "model.savedmodel"
    )
    self.model.save(actor_model_save_dir, \
                    save_format="tf")
    print(f"Actor model saved at:\
        {actor_model_save_dir}")
```

3. Next, we will add the `save_tflite` method to the `Actor` class to save and export the Actor model in TFLite format:

```python
def save_tflite(self, model_dir: str, version: \
int = 1):
    """Save/Export Actor model in TensorFlow Lite
    format"""
    actor_model_save_dir = os.path.join(model_dir,\
                            "actor", str(version))
    model_converter = \
        tf.lite.TFLiteConverter.from_keras_model(
                                    self.model)
    # Convert model to TFLite Flatbuffer
    tflite_model = model_converter.convert()
    # Save the model to disk/persistent-storage
    if not os.path.exists(actor_model_save_dir):
        os.makedirs(actor_model_save_dir)
    actor_model_file_name = \
        os.path.join(actor_model_save_dir,
                    "model.tflite")
    with open(actor_model_file_name, "wb") as \
    model_file:
        model_file.write(tflite_model)
```

```
        print(f"Actor model saved in TFLite format at:\
            {actor_model_file_name}")
```

4. Now, let's implement the save_h5 method and add it to the Actor class to save and export the Actor model in HDF5 format:

```
def save_h5(self, model_dir: str, version: int = 1):
    actor_model_save_path = os.path.join(
        model_dir, "actor", str(version), "model.h5"
    )
    self.model.save(actor_model_save_path, \
                    save_format="h5")
    print(f"Actor model saved at:\
            {actor_model_save_path}")
```

5. Next, we will add the save_tfjs method to the Actor class to save and export the Actor model in TF.js format:

```
def save_tfjs(self, model_dir: str, version: \
int = 1):
    """Save/Export Actor model in TensorFlow.js
    supported format"""
    actor_model_save_dir = os.path.join(
        model_dir, "actor", str(version), \
        "model.tfjs"
    )
    tfjs.converters.save_keras_model(self.model, \
                                actor_model_save_dir)
    print(f"Actor model saved in TF.js format at:\
            {actor_model_save_dir}")
```

6. As the final variant, we will add the `save_onnx` method to the `Actor` class to save and export the Actor model in ONNX format:

```
def save_onnx(self, model_dir: str, version: \
int = 1):
    """Save/Export Actor model in ONNX format"""
    actor_model_save_path = os.path.join(
        model_dir, "actor", str(version), \
        "model.onnx"
    )
    onnx_model = keras2onnx.convert_keras(
                    self.model, self.model.name)
    keras2onnx.save_model(onnx_model, \
                    actor_model_save_path)
    print(f"Actor model saved in ONNX format at:\
        {actor_model_save_path}")
```

7. That completes the save/export methods for the `Actor` class! In a similar way, let's add the `save` methods to the `Critic` class for completeness. Starting with the `save` method, and then the other methods in the later steps:

```
def save(self, model_dir: str, version: int = 1):
    critic_model_save_dir = os.path.join(
        model_dir, "critic", str(version), \
        "model.savedmodel"
    )
    self.model.save(critic_model_save_dir, \
                    save_format="tf")
    print(f"Critic model saved at:\
        {critic_model_save_dir}")
```

8. The next method in the sequence is the `save_tflite` method to save and export the Critic model in TFLite format:

```python
def save_tflite(self, model_dir: str, version: \
int = 1):
    """Save/Export Critic model in TensorFlow Lite
    format"""
    critic_model_save_dir = os.path.join(model_dir,\
                        "critic", str(version))
    model_converter = \
        tf.lite.TFLiteConverter.from_keras_model(
                                    self.model)
    # Convert model to TFLite Flatbuffer
    tflite_model = model_converter.convert()
    # Save the model to disk/persistent-storage
    if not os.path.exists(critic_model_save_dir):
        os.makedirs(critic_model_save_dir)
    critic_model_file_name = \
        os.path.join(critic_model_save_dir,
                "model.tflite")
    with open(critic_model_file_name, "wb") as \
    model_file:
        model_file.write(tflite_model)
    print(f"Critic model saved in TFLite format at:\
        {critic_model_file_name}")
```

9. Let's implement the `save_h5` add to the `Critic` class to save and export the Critic model in HDF5 format:

```python
def save_h5(self, model_dir: str, version: int = 1):
    critic_model_save_dir = os.path.join(
        model_dir, "critic", str(version), "model.h5"
    )
    self.model.save(critic_model_save_dir, \
                    save_format="h5")
    print(f"Critic model saved at:\
        {critic_model_save_dir}")
```

10. Next, we will add the `save_tfjs` method to the `Critic` class to save and export the Critic model in TF.js format:

```python
def save_tfjs(self, model_dir: str, version: \
int = 1):
    """Save/Export Critic model in TensorFlow.js
    supported format"""
    critic_model_save_dir = os.path.join(
        model_dir, "critic", str(version), \
        "model.tfjs"
    )
    tfjs.converters.save_keras_model(self.model,\
                            critic_model_save_dir)
    print(f"Critic model saved TF.js format at:\
        {critic_model_save_dir}")
```

11. The final variant is the `save_onnx` method, which saves and exports the Critic model in ONNX format:

```python
def save_onnx(self, model_dir: str, version: \
int = 1):
    """Save/Export Critic model in ONNX format"""
    critic_model_save_path = os.path.join(
        model_dir, "critic", str(version), \
        "model.onnx"
```

```
        )
        onnx_model = keras2onnx.convert_keras(self.model,
                                  self.model.name)
        keras2onnx.save_model(onnx_model, \
                        critic_model_save_path)
        print(f"Critic model saved in ONNX format at:\
            {critic_model_save_path}")
```

12. That completes the save/export method additions to our agent's `Critic` class. We now can add the corresponding `save` methods to the `Agent` class that will simply call the corresponding `save` methods on the Actor and Critic objects. Let's complete the implementation in the following two steps:

```
    def save(self, model_dir: str, version: int = 1):
        self.actor.save(model_dir, version)
        self.critic.save(model_dir, version)


    def save_tflite(self, model_dir: str, version: \
    int = 1):
        # Make sure `toco_from_protos binary` is on
        # system's PATH to avoid TFLite ConverterError
        toco_bin_dir = os.path.dirname(sys.executable)
        if not toco_bin_dir in os.environ["PATH"]:
            os.environ["PATH"] += os.pathsep + \
                            toco_bin_dir
        print(f"Saving Agent model (TFLite) to:\
            {model_dir}\n")
        self.actor.save_tflite(model_dir, version)
        self.critic.save_tflite(model_dir, version)
```

13. The remaining methods on the `PPOAgent` class are straightforward as well:

```python
def save_h5(self, model_dir: str, version: int = 1):
    print(f"Saving Agent model (HDF5) to:\
            {model_dir}\n")
    self.actor.save_h5(model_dir, version)
    self.critic.save_h5(model_dir, version)

def save_tfjs(self, model_dir: str, version: \
int = 1):
    print(f"Saving Agent model (TF.js) to:\
            {model_dir}\n")
    self.actor.save_tfjs(model_dir, version)
    self.critic.save_tfjs(model_dir, version)

def save_onnx(self, model_dir: str, version: \
int = 1):
    print(f"Saving Agent model (ONNX) to:\
            {model_dir}\n")
    self.actor.save_onnx(model_dir, version)
    self.critic.save_onnx(model_dir, version)
```

14. That completes our implementation for the `Agent` class! We are now ready to run the script to build, train, and export the Deep RL agent model! Let's implement the `main` function and call all the `save` methods that we have implemented in the previous steps:

```python
if __name__ == "__main__":
    env_name = args.env
    env = gym.make(env_name)
    agent = PPOAgent(env)
    agent.train(max_episodes=1)
    # Model saving
    model_dir = "trained_models"
    agent_name = f"PPO_{env_name}"
    agent_version = 1
    agent_model_path = os.path.join(model_dir, \
```

```
                                        agent_name)
agent.save_onnx(agent_model_path, agent_version)
agent.save_h5(agent_model_path, agent_version)
agent.save_tfjs(agent_model_path, agent_version)
agent.save_tflite(agent_model_path, agent_version)
```

15. It's time to execute our script! Please pull the latest copy of the recipe from the book's code repository and just run it! By default, the script will train the agent for one episode, save the agent models, and export the model in various formats ready for deployment. Once the script finishes, you will see the exported models with the directory structure and contents similar to the one shown in the following figure:

Figure 9.18 – PPO Deep RL agent model exported to various formats for deployment

That completes our final recipe for this chapter! Let's quickly revisit some of the key items we covered in this recipe in the following section.

How it works...

We first set TensorFlow Keras's backend to use float32 as the default representation for float values. This is because, otherwise, TensorFlow would use the default float64 representation, which is not supported by TensorFlow Lite (for performance reasons) as it is targeted towards running on embedded and mobile devices.

In this recipe, we used a PPO agent with Actor and Critic networks that expect image observations and produce actions in discrete space, designed for RL environments such as the procedurally generated procgen environment from OpenAI. You can swap the agent code with the PPO implementations from one of the earlier chapters that use different state/observation spaces and action spaces depending on your need/application. You could also replace the agent with a different agent algorithm altogether.

We discussed several approaches to save and export your agent models, leveraging the whole suite of tools and libraries offered by the TensorFlow 2.x ecosystem. A summary of the various export options that we implemented as part of this recipe is provided in the following figure:

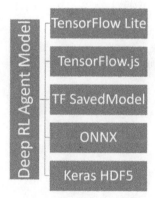

Figure 9.19 – Summary of various RL agent model export options discussed in this recipe

That concludes this recipe, the chapter, and – more dramatically – the book! We covered a lot of different topics in this cookbook to leverage the TensorFlow 2.x framework and the ecosystem of tools and libraries built around it to build RL building blocks, environments, algorithms, agents, and applications. I hope you had an exciting journey with the book.

I can't wait to see what you build/cook with the recipes we discussed in the book. I will look forward to hearing about your journey with the book on the discussion page at https://github.com/PacktPublishing/Tensorflow-2-Reinforcement-Learning-Cookbook/discussions.

Looking forward to getting in touch with you on the discussion/issues page. All the best for your future endeavors!

Other Books You May Enjoy

If you enjoyed this book, you may be interested in these other books by Packt:

Mastering Reinforcement Learning with Python

Enes Bilgin

ISBN: 978-1-83864-414-7

- Model and solve complex sequential decision-making problems using RL
- Develop a solid understanding of how state-of-the-art RL methods work
- Use Python and TensorFlow to code RL algorithms from scratch
- Parallelize and scale up your RL implementations using Ray's RLlib package
- Get in-depth knowledge of a wide variety of RL topics
- Understand the trade-offs between different RL approaches
- Discover and address the challenges of implementing RL in the real world

Deep Reinforcement Learning with Python - Second Edition

Sudharsan Ravichandiran

ISBN: 978-1-83921-068-6

- Understand core RL concepts including the methodologies, math, and code

- Train an agent to solve Blackjack, FrozenLake, and many other problems using OpenAI Gym

- Train an agent to play Ms Pac-Man using a Deep Q Network

- Learn policy-based, value-based, and actor-critic methods

- Master the math behind DDPG, TD3, TRPO, PPO, and many others

- Explore new avenues such as the distributional RL, meta RL, and inverse RL

- Use Stable Baselines to train an agent to walk and play Atari games

Leave a review - let other readers know what you think

Please share your thoughts on this book with others by leaving a review on the site that you bought it from. If you purchased the book from Amazon, please leave us an honest review on this book's Amazon page. This is vital so that other potential readers can see and use your unbiased opinion to make purchasing decisions, we can understand what our customers think about our products, and our authors can see your feedback on the title that they have worked with Packt to create. It will only take a few minutes of your time, but is valuable to other potential customers, our authors, and Packt. Thank you!

Index